THE
INDIVIDUALIZED
CORPORATION

THE
INDIVIDUALIZED
CORPORATION

A Fundamentally New Approach to Management

GREAT COMPANIES ARE DEFINED BY
PURPOSE, PROCESS, AND PEOPLE

**SUMANTRA GHOSHAL AND
CHRISTOPHER A. BARTLETT**

HarperBusiness
A Division of HarperCollins*Publishers*

HarperCollins books may be purchased for educational, business, or sales promotional use. For information please write: Special Markets Department, HarperCollins Publishers, Inc., 10 East 53rd Street, New York, NY 10022.

FIRST EDITION

Designed by Elina D. Nudelman

Library of Congress Cataloging-in-Publication Data

Ghoshal, Sumantra.
    The individualized corporation : a fundamentally new approach to management / Sumantra Ghoshal and Christopher A. Bartlett.
        p.   cm.
    Includes bibliographical references and index.
    ISBN 0-88730-806-6
    1. Decentralization in management.   2. Employee empowerment.
I. Bartlett, Christopher A., 1943–   .   II. Title.
HD50.G47   1997
658.4'02—dc21   97-18982

97 98 99 00 01 ❖/RRD 10 9 8 7 6 5 4 3 2 1

*To Aparna Ghoshal and Margot Stewart, our mothers,
with love and appreciation*

Contents

Acknowledgments: Shipmates on a Voyage of Discovery

This book represents the log of a voyage of discovery that began almost fourteen years ago, when we were undertaking the research for our previous book, *Managing Across Borders*. It was only when we were several years into our explorations that we realized that we were witnessing nothing less than the unfolding of the most profound change in management in a lifetime. Stimulated by this unique opportunity and supported by our respective institutions, we have been fortunate to have developed an unusually satisfying, rewarding, and productive relationship that has resulted in scores of case studies, dozens of articles, several classroom texts, and now two major research books. Over the past fourteen years, our working partnership has developed to the point where we can (and often do) complete each other's thoughts and finish each other's sentences. This is a truly equal partnership and we have long since given up trying to disentangle the origin of ideas or the authorship of papers—a recognition of which is reflected in our practice of alternating the first-listed author in successive articles and books. So, at the risk of self-indulgence, we first acknowledge the debt we owe to each other for the patience, trust, and commitment that was necessary to cement a solid working relationship into a truly committed partnership. In retrospect, it is clear to us that the output of the joint work has been more than a pile of written documents; it has been a genuine mutual respect and a strong friendship that will endure long after the reprints and footnote citations have faded from memory.

The next acknowledgment must be to the more than five hundred managers we have interviewed in over forty companies that have collaborated with our projects since the early 1980s. By allowing us see the world through their eyes, these managers have been the navigators, pilots, and helmsmen who have helped us understand the new and unfamiliar territory we were trying to chart and describe. To the extent we have been able to provide any understanding of the *terra incognita* of the emerging corporate form and the new management model, it is due to this experienced group. Although too numerous to acknowledge individually, they know who they are and will recognize their contributions as they read the pages that follow. To all of you, our deepest thanks for your insights, openness, and patience.

On any voyage, there are many who work "below deck"— those whose labors are often invisible, yet whose contributions are immense. Our thanks go to the many doctoral students and research associates who gathered data, wrote up company visits, and helped sharpen our understanding on various aspects of this project, including Mary Ackenhusen, Rob Lightfoot, Takia Mahmood, Afroze Mohammed, Peter Moran, and Ashish Nanda. For helping to bring meaning and clarity out of our often fuzzy prose, we also extend our gratitude to the many talented journal and manuscript editors who shaped our work, including Steven Bennett, Simon Caulkin, Tom Richman, Nan Stone, and the numerous anonymous reviewers of our articles in academic journals, and particularly our editors at HarperBusiness, Laureen Rowland and Kirsten Sandberg.

Like any long voyage of discovery, our venture required substantial patronage and backing, and we must also acknowledge the vital role played by the institutions to which we belong. Harvard Business School, INSEAD, and London Business School have been generous not only in their funding of this project but also in their flexibility in providing us the time to work on the research and writing. Furthermore, our colleagues at these institutions have been equally generous with their time and thought in reading papers, offering criticism, suggestions, and ideas that have greatly influenced our work. The number who have made such contributions are too numerous to list here, but as our friends, as well as our colleagues, they know who they are.

Finally, we reserve our greatest debt of gratitude for our families, whose encouragement and support have helped us through the many difficult parts of this long journey. To Susmita, Ananda, and Siddhartha and to Barbara, Nick, Liz, and Andrew, we want you to know how important your love, understanding, and unquestioning support was to us through what must have seemed like an interminable project. You sustained us through it all, and for that we offer you our heartfelt thanks and love.

Part 1

INTRODUCTION: BIRTH OF A NEW CORPORATE MODEL

The Rediscovery of Management: From Organization Man to Individualized Corporation

1

In 1682, the English astronomer Sir Edmund Halley had the good fortune to be at the right place at the right time. His observations on the spectacular comet that now bears his name helped earn him the prestigious title of Astronomer Royal. More important to this professional scientist was the fact that this fortuitous opportunity inspired a flurry of other research activity that led to new and important discoveries about the nature of our universe.

While hardly on the same grand scale as Sir Edmund, we, too, have been fortunate to have had a front-row seat at a once-in-a-lifetime event—the collapse of an outmoded corporate form and the emergence of a new management model that we believe will propel today's companies well into the twenty-first century. At the heart of the emerging model lie not only some very different organizational practices and processes but also a fundamentally different management philosophy. In this book, we will describe this new management model and illustrate how some of the pioneers of this new approach have implemented it in their own companies.

But in order to fully understand where we are going it is important to recognize and acknowledge where we have been. So before we launch into our voyage of discovery, let us take you on a brief trip through the incredibly short history of the modern corporation—a history whose distinct stages are, rather remarkably, punctuated by the last few visits of Halley's comet.

OF COMETS AND CORPORATIONS

When it blazed past our planet in 1835, the comet heralded the birth of a new corporate form emerging simultaneously in Britain, the United States, France, and Germany. The concept of limited liability led to a proliferation of new corporations eager to meet the demands triggered by the production and transportation booms that had followed in the wake of the Industrial Revolution. In contrast to an earlier generation of specialist traders, producers, wholesalers, and merchants, many of these new companies began to develop as vertically integrated organizations that required a higher order of management skills to coordinate their multifunctional operations.

When Halley's comet returned from its next tour of the solar system seventy-six years later, the well-established functional organization was showing early signs of reaching its limits, and a new corporate model was emerging. Responding to the simultaneous demands of increasing scale (triggered by the new mass-production technologies), and increasing scope (encouraged by diversification opportunities for leveraging existing capabilities into new markets), a few pioneering leaders like Alfred Sloan at General Motors and Pierre Du Pont at Du Pont Corporation began experimenting with the revolutionary multidivisional corporate form. This innovation, in turn, gave birth to an entirely new model of "professional management"—one that has become deeply embedded in practice, spread by consultants, and legitimized in textbooks and business school cases over the past three-quarters of a century.

With the era of mass production came the new and powerful philosophy of scientific management espoused by Frederick Winslow Taylor, an engineer and inventor who penned the highly influential *Principles of Scientific Management* (1911), based on his experience at Midvale Steel and Bethlehem Steel. Built on a belief that any task could be rigorously studied through time-and-motion analysis, broken into discrete activities, and executed by trained specialists, Taylorism not only resulted in a redefinition in the jobs of employees on the factory floor but also caused a revolution in the responsibilities of those who supervised them. Management's critical role became one of motivating and controlling workers in their tightly defined job descriptions, then ensuring effective coordination across these highly specialized activities.

At the same time, the diversification opportunities facilitated by the new multidivisional structure also had a powerful impact on management roles. In an environment in which capital had become the critical strategic resource, planning became a central part of modern management practice. With multiple divisional managers bidding for limited financial resources, companies began to develop sophisticated capital budgeting systems and strategic planning processes to guide the investments. To manage these processes and the related control systems that accompanied them, headquarters groups were expanded to ensure the quality of the information and to analyze it. As staffs' influence grew, so, too, did the policies and procedures that became their primary means of influencing operations.

But the new corporate model reflected more than just a new, complex structural framework and a more disciplined management approach. In this environment of control and coordination and of proliferating plans and policies, a new and different relationship started to develop between companies and their employees. Management's increasing preoccupation with the task of allocating capital and measuring its effective utilization led many to think of the company's employees less as valuable human resources and more as replaceable parts in an efficient production process. The difficulty was that unlike other inputs, people came with widely divergent capabilities and dispositions that made their behavior less predictable. As Henry Ford is reported to have said in exasperation: "When all I want is a good pair of hands, unfortunately I must take them with a person attached!" Most of the complex new structures and systems overlaid with increasingly sophisticated policies and procedures were designed to minimize individual idiosyncrasies and make people "as predictable and controllable as the capital resources they must manage," as Harold Geneen put it when he was ITT's legendary control-oriented CEO in the 1970s.

This divisionalized structure with its associated professional management approach spread rapidly, fueled by the postwar boom that created market opportunities that exceeded companies' abilities to fund them and fanned by consulting firms such as McKinsey and Company that exported the model and made it as common in Europe as it had become in the United States. In an era of economic growth and corporate expansion, most studies focused on the powerful new strategy-structure linkage, particularly the way in which the innovative multidivisional organization model facilitated rapid product and geographic diversification. But among the few observers who became interested in the impact the new corporate model was having internally on human behavior, nobody was more insightful or more influential than sociologist William Whyte, whose book *The Organization Man* became a major best-seller. What Whyte powerfully articulated was the way in which modern corporations had subjugated individual initiative and creativity to the perceived greater need for consistency and control.

The Passing of "Organization Man"

In 1986, Halley's comet once again passed by the third planet from the sun, leaving nothing but a vanishing trail of ice in its wake. By this time, once great companies—including such pioneers of the previous corporate revolution as General Motors and Du Pont—were leaving their own trail of abandoned structures, discarded controls, and downsized activities as they attempted to break out of the strategic inertia and organizational lethargy that had begun to envelop them.

Yet through all the adjustments, redesigns, and change programs, the deeply embedded assumptions that had created and supported the notion of the "Organization Man" remained largely unchallenged. Like most well-established and strongly held beliefs, this management paradigm became highly resistant to change, not least because to do so would question the established role and authority of the managers themselves. Just as the clergy ridiculed and then punished Copernicus and later Galileo for proposing a blasphemous heliocentric theory that challenged the supremacy of the earth's position, so, too, did managers resist and reject new management ideas that questioned their authority and control by building a new corporate model on a more liberating set of assumptions about human capability and individual motivation.

By the mid-1980s, however, a few pioneering managers began to question many of the fundamental management practices and corporate beliefs on which they had been raised and through which they had made their careers. One such visionary was Jack Welch, who became a role model for hundreds and perhaps thousands of CEOs worldwide as he gradually dismantled much of the classic corporate model that his predecessors had built at General Electric. Reflecting on his own learning, Welch evolved from a traditional hard-edged authoritarian who had earned his nickname of "Neutron Jack" to a more people-sensitive manager who understood the importance of treating his employees as sources of initiative, energy, and creativity rather than just as controllable costs. Said Welch:

> The talents of our people are greatly underestimated and their skills are underutilized. Our biggest task is to fundamen-

tally redefine our relationship with our employees. The objective is to build a place where people have the freedom to be creative, where they feel a real sense of accomplishment—a place that brings out the best in everybody.

On the other side of the Atlantic, another visionary leader was equally concerned about the poor job that his company was doing in capturing the energy, imagination, and commitment of its employees. Percy Barnevik, CEO of Asea Brown Boveri (ABB), felt that managers spent too much time trying to squeeze the last percentage point of productivity out of their capital assets while ignoring the huge untapped potential of their human resources:

> There is tremendous unused potential in our people. Our organizations ensure they only use 5 to 10 percent of their abilities at work. Outside of work they engage the other 90 to 95 percent to run their households, lead a Boy Scout troop, or build a summer home. We have to learn how to recognize and employ that untapped ability that each individual brings to work every day.

Outstanding leaders like Welch and Barnevik are today's Alfred Sloan and Pierre Du Pont. In a dynamic global environment in which competition was increasingly service-based and knowledge-intensive, they recognized that human creativity and individual initiative were far more important as sources of competitive advantage than homogeneity and conformity. Rather than forcing employees into a corporate mold defined by policies, systems, and constraints, they appeared to see the core task in almost the exact opposite terms. Their challenge was not to force employees to fit the corporate model of the "Organization Man" but to build an organization flexible enough to exploit the idiosyncratic knowledge and unique skills of each individual employee. It is a model we termed the "Individualized Corporation."

THIRD-GENERATION STRATEGIES, SECOND-GENERATION ORGANIZATIONS, FIRST-GENERATION MANAGERS

Like Welch and Barnevik, many of today's top corporate executives have come to recognize that they are leading their companies through a unique period of history. The last visit of Halley's comet coincided with the convergence of several important environmental forces that were forcing most companies not merely to adjust or adapt as they had in the past but to confront the need for major transformational change. Among the most powerful of these external forces were the exploding opportunities opened up by the globalizing economy, the technological demands of shortening product life cycles and shifting technology platforms, the competitive imperatives created by converging industry boundaries and expanding alliance partnerships, the structural realignment dictated by large-scale deregulation and shifts in the location of strategic assets, and the internal learning capabilities required to succeed in a knowledge-intensive environment in the fast-emerging information age.

The more managers understood about these new demands, the more they realized that their historic focus on adjusting the strategy-structure linkage was only part of a much more profound metamorphic change their companies had to make. Indeed, as companies adopted the new, multitiered, divisionalized structures and sophisticated capital-allocating processes we described, most were stumbling on a similar constraint: Their managers were simply unable to adapt to the demands being placed on them by the new organization. Not only were the new roles and relationships more complex and less clear, but the existing employees' skills and experiences were often unequal to the needs of the new jobs. One manager captured the situation accurately when he suggested that his company "was trying to implement third-generation strategies through second-generation organizations run by first-generation managers."

Beyond Corporate Tune-Ups

The fast-changing demands of an ever more complex operating environment—the forces driving the need for third-generation strategies—created an urgent requirement for new organizational paradigms and innovative management models. It was a fertile breeding ground for the new ideas of academics and consultants alike, and managers soon found themselves trying to identify a laser-focused strategic intent while simultaneously developing a broad corporate vision; they labored to structure tight network linkages within structures they had framed as loosely coupled virtual organizations; and they attempted to reengineer complex work processes while also empowering their frontline managers.

By the mid-1990s, the managers of most large companies felt that their tool kits were overflowing and their apprenticeships had been served. Yet most remained confused about what it was they were creating. Worse still, the outcomes were disappointing: Many of the change initiatives had proven unsuccessful, and even where they had seemed to pay off, they had not bridged the widening gap between their companies' strategic imperatives and their organizational and managerial capabilities. The generational gaps persisted.

The extent of the problem was documented in a 1994 study conducted jointly by The Planning Forum and the consulting firm Bain and Company. They found that over the previous five-year period, the average company had committed itself to 11.8 of the 25 popular tools and concepts listed on the questionnaire—from corporate visioning and empowerment to reengineering and activity-based costing. And this search for solutions seemed to be accelerating. Having employed more than two new tools or techniques annually for the previous five years, the typical company was planning to adopt another 3.7 in the coming twelve months. Yet despite this frenzy of activity, the overall satisfaction level with most of the tools was quite low, and the authors could find no correlation between the number of tools a company employed and its satisfaction with its performance.

The problem with most of these tools and techniques is not that they are ineffective, for indeed most are quite valuable. The reason for their limited success is that they are inadequate for the huge transformational task at hand. As the Bain survey indicated, most companies are adopting them as a series of random programmatic initiatives when what is needed is a more fundamental systemic change. As Welch, Barnevik, and other pioneers have recognized, a company cannot maintain its momentum and effectiveness just by implementing a series of tune-ups on a corporate engine that needs replacing.

Between 1990 and 1996 we spent thousands of hours interviewing hundreds of managers in twenty companies that we felt could help us gain a deeper insight into the nature of the new corporate engine and how it could be built and installed. While recognizing the power of studying the missteps and stumbles of former corporate icons, we chose to focus most of our attention on companies that were defining the emerging corporate form. Not that we were trying to find some idealized model that we could set up as an objective for others to follow. Rather, we sought to gain insight into a variety of successful new forms and transformational processes so we could stand back and provide useful generalizations.

Some of the companies, like GE and ABB, were in the midst of an often painful transformational process; others such as Intel and IKEA were born in the new era and had developed many of the new characteristics from scratch; and a third group including 3M and Kao Corporation seemed to have avoided the most debilitating features of the traditional management approach by evolving to a different corporate model over many decades. In each case, however, they focused us on interesting questions that helped us piece together a better understanding of what was emerging: a radically different corporate form, supported by a revolutionary management philosophy. For example,

- How was ABB able to integrate a disparate collection of second- and third-tier companies into a highly competitive organization that is now the clear industry leader in the global power equipment business?

- How has Intel not only survived the competitive blood-bath that wiped out most of its competitors but gone on to dominate the semiconductor industry through its ability to constantly obsolete its own leading products?

- How did the Kao Corporation leverage the knowledge of its mature soap and detergent business into a diversified portfolio of successful businesses, from paper products to floppy disks, and, in the process, build a reputation as one of Japan's most innovative companies?

Pursuing the answers to these and dozens of similar questions was an exhilarating yet humbling experience. Much of what we learned in the field challenged the basic assumptions and core beliefs that we and our colleagues had been teaching for decades. More important, what we observed ran counter to the mainstream organizational philosophies and management practices that guided the operations of the vast majority of companies worldwide. As a small but growing band of pioneering companies and innovative managers were joined by a swelling contingent of others, there was a clear need for charts, no matter how crude, to help the voyagers navigate their crossing from a corporate model dependent on its ability to shape and control the "Organization Man" to one committed to the notion of becoming an "Individualized Corporation."

A Blueprint for Change

At all levels of the organization, managers are overwhelmed and confused. They certainly don't need another rallying cry to transform their organizations; they need to know "Into what?" Neither do they need more slogans about reinventing management; they want to know how. It was this frustration we sensed in companies throughout the world that provided the focus and impetus for our ambitious six-year project.

If our motivation for undertaking the research was to obtain a firsthand understanding of the transformation process currently sweeping through companies worldwide, then our goal in writing this book is to provide practicing managers with an overview of the new corporate model that is emerging. We believe that only after answering questions relating to the why and what of the ongoing corporate revolution can a manager really understand the how. Therefore, rather than focus on the artifacts of the problems, we want to help managers understand their root causes. And instead of adding to the proliferation of prescriptive tools and normative techniques, we have developed a set of integrated ideas and a conceptual framework that we believe provides them with a mental map of the new corporate terrain.

Within this broader objective, we will also describe how the managers of some of the companies we studied have applied these concepts and implemented the change from a traditional corporate paradigm to the emerging model we characterize as the Individualized Corporation. While not offering either a universal solution or a quick-fix prescription, we believe we can provide managers with a perspective that will allow them to make sense of the revolution in which they find themselves, as well as suggest some ideas to help them manage successfully through it.

In this introductory section, we present the concept of the Individualized Corporation. Having framed the historical context and described the research motivation in this chapter, we will use the next to provide a particular example of how the transformational changes at one company affected the activities, motivation, and eventually the performance of one manager who had spent his whole career as a classic Organization Man.

In part 2 of the book, we describe the organizational characteristics required to develop three core capabilities that distinguish the Individualized Corporation. The first is its ability to inspire individual creativity and initiative in all its people, built on a fundamental faith in individuals. In chapter 3, we will illustrate how companies like 3M, one of the most admired corporations in the United States, and ISS, the Copenhagen-based global leader in the office-cleaning business, have developed this ability, and will draw some general lessons from their organizational and management practices to suggest how others may learn from them. The second core characteristic of the Individualized Corporation is its ability to link and leverage pockets of entrepreneurial activity and individual expertise by building an integrated process of organizational learning. In chapter 4, we will describe this process, drawing on our observations at McKinsey and Company, the world's premier strategy consulting firm, and Skandia, the Scandinavian insurance company, to illustrate how managers can build an organizational learning capability in their own companies. Chapter 5 will focus on the third core feature of the new corporate model, its ability to continuously renew itself, and the Kao Corporation, one of the most creative companies in Japan, and Intel, the global leader in the semiconductor business, will serve as examples of companies that have been able to develop such a capability.

Part 3 of the book explores the managerial implications of these new organizational characteristics and focuses on how companies can go about building and managing the Individualized Corporation. Using the transformation of the semiconductor business by Philips, the European electronics giant, as an illustration, we will argue in chapter 6 that at the heart of the new corporate model is a carefully constructed organizational context that frames and supports a very different set of human motivations and behaviors. Then, in chapter 7, we will propose a concept of the new organization not as a hierarchy of tasks but as a portfolio of processes, and use Asea Brown Boveri (ABB), perhaps the most admired company in Europe in the mid-1990s, as a classic example of a company built this way. This different conceptualization of the organization has radical implications for the roles of frontline, senior, and top-level managers of a company, and for the relationships that are needed among them to tie these new roles together into a high-performing management system. In chapter 8 we will examine the ways in which these redefined roles influence the personal competencies required of those who will manage such organizations, and how companies like ISS, 3M, and McKinsey have helped their managers develop them. Part 3 concludes with a chapter describing a transformation process in terms rather different from the restructuring and reengineering approaches that have dominated recent thinking. Tracking the changes at General Electric in chapter 9, we will outline a three-phased process of corporate renewal, showing how changes in the strategic and structural "hardware" need to be matched with changes in the contextual and behavioral "software" within each of the phases in order to build the management capabilities of the Individualized Corporation.

Finally, in part 4 of the book, we step back from the details of defining the characteristics of the Individualized Corporation and describing how to build and manage it, to examine the more profound issue of a new managerial philosophy that underlies this new corporate model. Rather than accepting the economists' assumption of the company merely as an economic entity and of its goal as appropriating the greatest possible value from all its constituencies, we take a broader and more positive view. The philosophy we propose is grounded in the belief that the company, as one of the most important institutions of modern society, must serve as its key engine of progress by creating new value for all its constituencies. This new philosophy creates a very different moral contract among employees, companies, and society and, in chapter 10, we will describe how the Unipart Group of Companies in the U.K. have fundamentally changed their relationships with customers, suppliers, employees, and their local communities by implementing such a contractual change. Then, in chapter 11, we will describe and illustrate, using the case of IKEA, the Swedish furniture company, how this new contract is fundamentally redefining the role of corporate top management, away from its classic focus on strategy, structure, and systems that led to the creation of the Organization Man, to one built around purpose, process, and people that embodies the managerial philosophy of the Individualized Corporation.

Halley's comet will return to our corner of the solar system in the year 2062. When it does, we believe our successors will look back at the Individualized Corporation as the revolutionary new model that brought business organizations into the postindustrial, information-intensive age of knowledge. Like Halley himself, we hope our findings will prove interesting and useful and will stimulate further research and thinking about the revolutionary corporate form now emerging. As for the future, we can only speculate what the new management model may give rise to as the great comet and the course of business management both continue on their ineluctable and seemingly intertwined paths.

Rebirth of an Organization Man: One Manager's Rediscovery of Management

2

Among the hundreds of site visits we made in the course of our research, one had more influence on our thinking than all others. It was this visit that first planted in our minds the seed that later grew into the concept of the Individualized Corporation. It was a visit to Coral Springs in the United States to study ABB's U.S. relays business located there and to meet with Don Jans, the general manager of the unit.

We went to Coral Springs to study firsthand a rather remarkable turnaround story. Historically a part of Westinghouse's Transmission and Distribution business, the unit had a record of modest profitability and almost no growth. But after it had been acquired by ABB in 1989, its revenues had grown by more than 45 percent in four years, while its profitability had improved by 120 percent. On-time shipments had improved from 70 to 99 percent, cycle time had been cut by 70 percent, and inventories had been slashed by 40 percent. Overall, a mature operation in a mature business had developed, almost overnight, the performance profile of a young growth company.

This might have been just another impressive but otherwise unremarkable turnaround—except for one thing. The transformation was accomplished by the same management team that had previously delivered the flat sales and break-even profitability. We had often heard from managers that talk about corporate renewal was fine, but what was needed was an overhaul of personnel: "You can't teach old dogs new tricks," they said. Don Jans was about as old a "dog" as you could get: He had spent thirty-two years in Westinghouse, the last three as the general manager of the relays business. Yet it was he who had driven the radical performance improvement of that same unit. Not only he but Joe Baker, his geographic boss in ABB's matrix organization, was also a Westinghouse veteran of thirty-nine years. How could the same people, managing the same business, achieve such radically different results with a change of corporate ownership?

Baker provided the answer:

> In Westinghouse, we recruited first-class people, did an outstanding job of management development, and then wasted all that investment by constraining them with a highly authoritarian structure. In ABB, we spent much of our first year thrashing out how we would work together. [ABB's senior management] spent a huge amount of their time in day-to-day intensive communication, taking the message to the frontline managers that they were responsible, that they needed to initiate, to question, and to challenge. In the end, it was this culture of delegated responsibility and intensive communication that made this organization work. It was an amazing change—I felt like I'd rediscovered management after thirty-nine years.

What Baker and Jans had discovered was the magic of the Individualized Corporation, built on the bedrock of individual initiative at all levels of the company. And, while ABB may be exceptional, it is by no means unique. In our study, we found Theo Buitendijk in ISS also "rediscovering management" as he moved from Exxon to ISS, the Copenhagen-based global cleaning services company. In 3M, we found the same spirit of individual initiative driving Andy Wong as he and his team developed a new computer privacy screen, creating yet another new business in this $14 billion industrial behemoth that continued to act like a high-tech start-up.

But Jans's story is the most compelling of all because it is the story of the rebirth of an Organization Man running the same business in the same organization. The message is, it is not always those on the front lines who lack initiative, creativity, and drive: It is more often management that holds them hostage in their corporate hierarchies. For all those managers who disown responsibility under the comfortable belief that "you can't teach old dogs new tricks," the metamorphosis of Jans and his team will raise unsettling doubts. That is why we begin our description of the Individualized Corporation by recounting the tale of one man's rediscovery of management.

PROFILE OF A CAREER MANAGER

Don Jans signed on with Westinghouse as a junior engineer in 1956, taking on an entry-level position in the Buffalo Motor division. Working his way up through a series of engineering, sales, and manufacturing positions, Jans was prepared for the first of three general management positions he would assume during the 1980s. But this was a difficult period in Westinghouse's history, and for more than a decade Jans had watched the company undergo a tumultuous series of changes, during which the executive suite seemed to have a revolving door.

In 1975, when Robert Kirby took the helm as CEO, Westinghouse had just suffered a $1 billion loss. To support its nuclear power business, the company had bet on the uranium futures market and lost. While it had signed long-term contracts with its utility customers to supply uranium at $9 per pound, the market price had soared to $40 per pound.

Kirby traced Westinghouse's problems to two root causes: over-diversification that had spread the company's strategic and managerial resources too thinly, and overdecentralization that had insulated top management from the true direction and performance of the businesses. To overcome the first problem, Kirby divested fifteen major businesses and withdrew from many of the company's activities outside the United States. To deal with the second concern, he implemented a comprehensive strategic-planning system at the corporate level to manage the company's portfolio of businesses in a more coordinated manner. On top of this, he institutionalized a system known as Vabastram (Value Based Strategic Management) developed by the corporate planning department to guide the company into growth businesses while restoring managerial and financial discipline in the operating divisions. The complicated planning model required business unit managers to measure their performance on an equity value, which corporate staff calculated for each of their businesses based on the cash flows projected from a range of alternative strategies. The system was designed to decide which businesses were to be retained in the Westinghouse portfolio, and to select among alternative strategies by implementing those that generated the highest value.

For several years, Kirby's strategy of restructuring operations and recentralizing control yielded good results as inefficient plants were closed, marginal businesses sold, and management attention was focused on cost control. Furthermore, Vabastram was a highly disciplined resource allocation tool that made no concessions to the company's traditional businesses, and like many employees, Jans was saddened as he saw the sale of old core businesses like the lighting, appliance, and motors business, where his own Westinghouse career had begun.

In this tumultuous organizational context Jans was offered his first general management job heading the company's underground distribution transformer business. Driven by the relentless demands of the new systems, he pushed hard on costs but found that after years of cutting, most of the juice had been squeezed from the division. If he was to meet his numbers, the only alternative was to boost prices. So in 1981, in the midst of the recession, he increased prices five times in a single year.

Seen as a manager who could respond to the need to cut and trim, Jans was then asked to head up a transformer components division newly created from the rationalized operations of five plants and personnel in eight locations. Once again he attacked the cost and revenue targets the company's new systems-driven management process demanded. After two years of battling to bring costs down by 15 percent, Jans found that the division's activities were pulled apart again in a fresh round of rationalization. Like many others, he found that his position had been eliminated.

By 1983, as Kirby handed over the reins to Douglas Danforth, the cost of starving the core businesses was beginning to take its toll. In the midst of a global recession, the financial performance of the business had slumped back almost to where it was when he had begun. Danforth attacked the challenge with energy and a new, more expansionist approach that initially provided some hope for frontline managers like Jans in Westinghouse's traditional businesses. Describing the company's situation as a "business gridlock" and ascribing many of its problems to overcontrol, Danforth began a new phase of decentralization. He eliminated a number of layers of corporate staff and greatly simplified the headquarters' review requirements for business units' strategic and financial plans.

Yet in an environment in which adversarial and competitive relationships had developed between line managers and the powerful corporate staff, the decentralization only increased the feeling of paralysis. In the Kirby era, at least the centralized bureaucracy had both the information and the power to take action. Now, with the decentralized organization trying to keep the headquarters in the dark about any bad news—and yet operating without the resources or the management norms to initiate strong initiatives—no one had the ability to do anything. As described by one Westinghouse veteran, Danforth's decentralization only succeeded in diluting the hard edge of performance accountability and increasing the rewards for political gamesmanship.

Having had his line job eliminated, Jans worked at corporate headquarters as assistant to the president of the Industries and International Group through most of this period. He reflected on the culture that had developed in the company and particularly among headquarters executives:

> Vabastram was basically a Wall Street play, but its main internal effect was to force managers to take a very short-term view in order to maximize their impact on stock value on a quarterly basis. For example, our international strategy focused more on export and licensing than foreign investment. As an operating manager, I became particularly troubled when we stopped being concerned about capitalizing on our strong competitive positions or defending our leading market shares. If the operating performance was down, Vabastram committees would meet and we would sell off assets to keep performance pumped up.

During the Danforth era, Westinghouse divested a whole slew of businesses and acquired fifty-five others in a five-year period. More than ever, managers recognized that the management philosophy was, as *Forbes* characterized it, "If a business performs, keep it. If not, dump it." And, as they all understood, "performance" meant earning more than your attributed cost of capital quarter by quarter.

In 1986, at the end of Danforth's tenure as CEO, Don Jans was offered an opportunity to return to a general management job heading the company's relays business based in Coral Springs, Florida. Anxious to return to line operations, he grabbed the chance. He certainly felt he was well prepared—his earlier experience running frontline businesses had proven him to be a strong operating manager, and his three-month stint at Harvard's Advanced Management Program had stimulated him to come up with lots of new ideas. It was a great opportunity for him to use his experience, exercise his initiative, and implement the many new ideas he had developed.

To Jans's frustration, however, he quickly discovered that he couldn't. Vabastram was still king, and the relays operation was part of the "mature" utilities business that was still in a downturn and therefore not driving Westinghouse's stock price up. In this environment, Jans soon found he had to radically revise his ideas for building the business. He described the environment in Westinghouse at the time:

> The company's leadership was increasingly drawn from managers with no experience in the core utilities businesses. They didn't understand those businesses and mostly they didn't like them because they had dominated investments for decades. As a result, they kept setting targets for us that we couldn't meet, so it became hard to earn a bonus and next to impossible to get capital.

It soon became clear to Jans that he was there to manage an endgame. Despite the appearance of new competitors offering solid-state microprocessor-based relays similar to those already in widespread use in other markets worldwide, Westinghouse held off investing in the development of new products, choosing instead to continue manufacturing its traditional electromechanical line. Even if the business wasn't sold off for performance reasons, it became increasingly clear that it would eventually die a natural death from technological obsolescence. In effect, Jans and his team were serving as the crew of the *Titanic*. The difference, of course, was that everyone at Westinghouse had a clear sense of the impending doom.

NEW LIFE FOR AN OLD BUSINESS

In 1989, the hour of reckoning appeared to have arrived. At that time, Asea Brown Boveri (ABB), the Swedish power equipment company, made an offer to acquire 45 percent of Westinghouse's transmission and distribution division, of which the relays business was a part. When the news of the deal made its way through the company grapevine, Jans and his team members began to update their résumés. Knowing that ABB already had its own more modern relays operation in the United States, they assumed that the old-time Westinghouse managers would be swept aside in the takeover. But, to their surprise, ABB invited most of the key people to stay on, even after purchasing the remaining 55 percent of the business the following year. Recalled Jans:

> The prevailing view when ABB acquired us was that we'd lost the war. We were resigned to the fact that the "occupying troops" would move in and we'd move out. But, they not only asked us to stay on, they gave us the opportunity to run the whole relays business—even the Allentown operation that was ABB's own facility in the United States.

Staying on, however, required Jans and his colleagues to manage in an environment like none they had ever seen at Westinghouse. More than simply making a transition from one employer to another, they felt they'd begun completely new careers, demanding fundamental changes in their business assumptions, organizational practices, and management styles. The context was completely different: At Westinghouse, Jans had five layers of management between himself and the CEO; at ABB, there were only two. At Westinghouse, he had been constantly frustrated by the bureaucracy imposed by a 3,000-person headquarters; at ABB, he had to adjust to the need for self-sufficiency in an organization with just 150 people in the corporate office. At Westinghouse, decisions had been top-down and shaped by political negotiations, whereas at ABB Jans found that those on the front line were expected to initiate much more, and that issues were decided on the basis of data and analysis.

The result, as Jans put it, was that management became an "exquisite challenge—a privilege." The whole focus of management's activities changed too, concentrating on ways to create opportunities and introduce innovations rather than finding ways to circumvent internal barriers or manipulate data to survive another review. To Jans and his team, the ABB acquisition of Westinghouse's relays business represented more than just a bridge to security; it became a pathway to fulfilling careers in an entirely new style of organization. And in that new corporate model the "organization man" in each of them was forced to become a persona of the past.

REDISCOVERING MANAGEMENT

At the simplest and most obvious level, one of the major shifts for Don Jans and his colleagues was the opportunity to work in a company that did not view power transmission and distribution as a mature or aging business. Looking beyond the difficulties the business faced at that time, Percy Barnevik, the CEO of ABB, was convinced that the decade-long sag in demand for power equipment would reverse itself as existing power plants in the industrial world became obsolete and as a large group of industrializing countries focused on building the infrastructure for their own entry into the twenty-first century. And he was ready to back up his convictions with substantial resources as well as his personal time and energy.

Barnevik knew, however, that he would have to do more than invest cash and enthusiasm for the new enterprise to succeed. To achieve his ambition of "conquering the globe" in the power industry, he would have to build a unique organization that could resolve some fundamental paradoxes. On the one hand, the new technologies and economies of scale necessary to meet the expected demand could be developed only by companies operating on a global scale, able to fully exploit the advantages of bigness. On the other hand, because of the high level of government ownership and control over utilities, companies with a strong national presence and with the flexibility and agility of a small business would garner most of the new orders. The real requirement of success, then, was not just resources or strategic brilliance at the top of the company but a broad-based organizational capability embedded deep into the corporate ranks. As he described in his vision for ABB:

> We have to be global and local, big and small, radically decentralized with central reporting and control. Once we resolve these dilemmas, we will achieve real organizational advantage.

Resolving these dilemmas required not only a new structure— ABB's highly publicized global matrix—but also a new organizational philosophy that would minimize internal competition, break down geographical, functional, and cultural barriers, and enable people to think and act entrepreneurially within the boundaries of the company. For that, the small, local, and radically decentralized elements had to become the new organization's foundation, its core; and the big, global, and central reporting and control characteristics had to be the overlays. As Barnevik constantly reminded his management team, it was an organization designed to encourage individual initiative and ensure personal responsibility:

> The only way to manage a large, complex company like ABB is to make it as simple and local as possible. The press may describe us as a $30 billion diversified global company, but we see ourselves as a portfolio of 1,200 companies, each with an average of two hundred employees. This is where the real work gets done, and these people need well-defined responsibilities, clear accountability, and maximum degrees of freedom to execute.

What this philosophy meant for Jans was nothing short of a total revolution of his own role within the company. From being an effective operational implementer working hard to be an effective part of a massive corporate machine, he was now cast in the role of an entrepreneurial initiator with full responsibility and accountability for the development of his own frontline company. As president of ABB Relays Inc., a separate legal entity created by ABB, he assumed full responsibility not only for his profit-and-loss statement but also for his balance sheet. This meant he had to focus on managing cash flow, paying dividends to the parent company, and making wise investments with his retained earnings, typically about 30 percent to 40 percent of total earnings. It also meant that he could borrow locally and that he inherited results from year to year through changes in equity. In short, he began seeing his job not simply as implementing the latest corporate program but as building a viable, enduring business.

Supporting and guiding him in these decisions, he had access to a seven-person steering committee that met three or four times a year and acted as a small local board for his frontline company. With membership drawn from ABB's global relays division (or worldwide business area in company terminology), the U.S. power transmission and distribution headquarters, and colleagues running related frontline companies within ABB, this steering committee became Jans's sounding board for new ideas (how to reorganize his unit, for example) and decision forum on key issues (such as approval for strategic plans and operating budgets).

In this new and challenging management framework, Jans was stretched beyond a preoccupation with his own operating unit. He was also invited to serve on the steering committees of the Canadian and Puerto Rican relays companies and ABB's closely related network controls company. As the head of one of the largest and most strategically important frontline relays companies, he had a seat on the worldwide relays business area board where global strategies and core policies were hashed out for the global businesses operations.

The radical decentralization of resources and responsibilities also penetrated deep into the formal structure. Like most company presidents, Jans quickly restructured his company into five profit centers, pushing responsibility and accountability down deeper into the organization. All of a sudden, managers who historically had thought of themselves primarily as engineers began to focus on market needs and became concerned about financial performance.

The philosophy of moving people and ideas beyond their traditional boundaries also touched staff groups, with many local specialists finding themselves on global functional councils through which they were expected to contribute their expertise to improving the company's worldwide performance. Councils on total quality, purchasing, human resource management, and other key staff responsibilities provided the linkage point that allowed functional experts in various companies worldwide to compare performance and transfer best practices on a routine basis.

Eventually, this culture of engaging people and stretching them to encourage individual initiative and a sharing of ideas penetrated to the front lines of the smallest operating unit. For example, as the relays business area board began working on a global strategy, it immediately reached down the management ranks of the local relays companies to identify nine high-potential technical administrative and marketing managers from Brazil, Germany, the United States, Finland, and Switzerland to develop a first-draft proposal. It was the ideas and proposals of this group that became the basis on which the $400 million relays business was built over the next five years.

While this structure of decentralized responsibility was a key element of what Don Jans and his fellow Westinghouse expatriates described as their "rediscovery of management," there was something else that was far less tangible—yet far more powerful—in the new context, something that helped them shed their roles as Organization Men and use their abilities in ways they could only have dreamt about at Westinghouse. It was a management model personified by the new leaders that redefined the very way they thought about their jobs.

WALKING THE TALK

From the very first meeting with ABB managers, Jans and his colleagues were swept away by the difference in management style. Nothing in their experience had prepared them for the kind of environment they found themselves in. Within weeks after the acquisition of the Westinghouse relays operation, Percy Barnevik and the then executive vice-president responsible for ABB's power transmission sector, Goran Lindahl, flew to the United States to express their confidence in the U.S.-based managers. The two senior ABB executives also sent a strong message that the acquisition would not follow the traditional takeover model in which the parent immediately establishes restrictive strategic and operating boundaries around the acquisition.

Moreover, Jans was amazed by the fact that Barnevik and Lindahl approached them as colleagues rather than as superiors. "These people weren't marketers, lawyers, or accountants," he recalled. "They were engineers who really understood the key business issues of the relays technology and marketplace. Unlike our old Westinghouse bosses, Barnevik and Lindahl also believed that the power transmission industry was about to enter an era of growth and were willing to invest in that future. What they wanted to know was: "How can we help you be a real contributor to this effort?"

As the relationship progressed, Barnevik and other top executives did not fade from the scene. Rather, to the delight of Jans and others on the front line, they seemed to become even more engaged, in ways that were both challenging and supportive. "We were constantly seeing the top guys in meetings and seminars," said Jans. "They showed hundreds of transparencies and could talk for hours about how the industry was developing, where ABB wanted to be, how it was going to get there, and so on. It was spellbinding—a real education."

Along with their clearly articulated vision about the future of the power industry, Barnevik and Lindahl also conveyed a strong sense of the company's core values. They talked, too, about the contribution that ABB could bring to the quality of life in the markets it served. As Barnevik explained it, ABB was not just in the business of selling power transmission equipment, it was in the business of improving living standards worldwide, bringing free enterprise and economic development to China and the countries of eastern Europe, and improving the environment by making smoke-belching, coal-burning power plants relics of the past. In doing so, he wanted to inspire people to connect with the company's broad mission in a very personal way; to see the company as the means by which they could have a personal impact on issues of major importance in the world.

More than just articulating an engaging vision and a set of values, Barnevik and his top team spent an enormous amount of time representing a management approach and operating style that reinforced the organization's belief that individual initiative and personal responsibility were at the heart of the company's philosophy. It was reflected in the respectful and thoughtful way they communicated within the organization and in their constant urging of individuals to question assumptions, propose solutions, and take action.

Nowhere was the new openness of communication more clearly evident than in the contrast between the strategic management processes at Westinghouse and at ABB. Where Vabastram was a top-down, staff-managed, financially driven model that focused managers on short-term operating performance under threat of divestment, ABB relied on an interactive, bottom-up/top-down process that was designed to engage managers at all levels in an ongoing dialogue about how to build and defend long-term sustainable competitive advantage. This approach was clearly demonstrated by the actions of Ulf Gundemark, head of the worldwide relays business area, who built his strategy on the study conclusions of his seven-person team of frontline operating managers. He said:

> I wanted to sweep aside a lot of the old assumptions about strategy that we inherited from the 1970s and 1980s—that it was defined primarily by top management, that it was communicated through numbered confidential copies, and that it was updated and reviewed annually, usually without challenging the underlying assumptions and objectives. I wanted strategy to become a process that involved all levels of management, was widely and freely communicated, and was constantly open to challenge.

The ability—indeed the requirement—for everyone to question and challenge was built into the matrix organizational framework, and reinforced by the management norms Barnevik and his top team continually emphasized. The company encouraged feedback and open debate, with the only qualification being that the differences be resolved and action taken. In such an environment, frontline managers like Jans were given greater freedom and more encouragement to take initiative than ever before.

When Jans decided he wanted to develop his company's ability to build electronic relays, he budgeted $1.5 million in new hires and product development expenses for the first year. Gundemark, his business area head, supported the idea, but Joe Baker, Jans's boss on the geographic axis of the ABB matrix and also an ex-Westinghouse man, resisted, citing budget shortfalls in the overall U.S. power transmission and distribution business.

Baker recalled Jans's response:

> Don really got mad and wrote me a strong letter. In Westinghouse, he probably would have been removed, but here we encourage people to kick back. I didn't like it, but I'm glad he did it. It led to some interesting discussions in his steering committee where I suggested to Ulf Gundemark that if he really wanted the R&D done, maybe he could support it out of Sweden for a year. In the end, Don got his program, and I kept my budget in line.

Events and outcomes such as these created an organizational atmosphere in which employees felt involved and engaged on an individual basis and at a very personal level. For the first time in his career, Don Jans felt as if he understood that the broader goals to which he was contributing were connected to a set of values that he found personally meaningful. At long last, he was fully involved in the decisions that shaped the world in which he operated.

ANOMALY OR MODEL?

At this point, it would be reasonable to ask how other companies can replicate what ABB achieved in transforming once stagnating businesses. Is this an anomaly that others could never replicate?

Does it hinge on charismatic leaders like Percy Barnevik? Were Jans and his team unusual in their ability to accept the profound changes necessary to manage in an individualized corporation?

After spending five years studying the major changes taking place in twenty very different companies in widely diverse businesses and from various countries worldwide, our answer to all three questions is a resounding "No!" While it is clear that the actual methods, tools, and processes for carrying on the transformation task will be different for different companies and adapted to the unique situations of each company, it is also equally clear to us that the kind of outcomes ABB achieved—both in changing people's behaviors and in neutralizing business performance—are attainable by others. While ABB may not be a *model* that every other company can adopt, it is an *example* of what most companies can achieve.

It is also worth emphasizing that ABB is not a perfect company, as anyone working within it would quickly acknowledge. Indeed, during one of our interviews with Don Jans he asked us to review an "organizational balance sheet" he had prepared to present at the next meeting of the worldwide relays business area board. On the assets side he listed a dozen attributes, such as clear vision and expectations from the top, integrating devices that facilitated collaborative management and mutual respect, and a strong ethic of "what, how, deadline." But in an equally long liabilities column, he highlighted concerns such as the tensions between business and country management, the barriers to technology sharing, and the distraction of time-consuming integration processes. Yet what made the challenge of managing in this manifestly imperfect company "an exhilarating experience" was in part reflected in Jans's preparation of the balance sheet that he presented for discussion by his bosses and colleagues. This was an environment in which individual initiative was celebrated, feedback and challenge expected, and the power of collaborative action an article of faith.

After observing many other companies developing the unwavering faith in the individual that was so clear in its absence at Westinghouse, we believe that any company fueled by such a commitment can tap into the same reservoir of entrepreneurship and collaboration that appeared spontaneously at Coral Springs—and, in doing so, achieve the same performance results for their business.

Part 2

FROM ORGANIZATION MAN TO INDIVIDUALIZED CORPORATION

Inspiring Individual Initiative: Building on a Belief in the Individual

3

Most companies look very different from the top than from the bottom. From the top, the CEO sees order, symmetry, and uniformity—a planned and precise instrument for step-by-step decomposition of the company's tasks and priorities. From the bottom, the hapless frontline manager sees a cloud of faceless controllers—a formless sponge that soaks up all his energy and time. The result, as described colorfully by General Electric's CEO, Jack Welch, is "an organization that has its face toward the CEO and its ass toward the customer."

In the last chapter, we described the way the company looks from the bottom, from the perspective of Don Jans's experiences at Westinghouse. Our purpose was not to single out Westinghouse for its poor management—in fact, during this period, Westinghouse leadership was frequently cited as being on the cutting edge in its application of new systems and programs. Rather, our objective was to provide an example of how modern management approaches and rational corporate models are creating an environment in which thousands of capable individuals are being crushed and constrained by the very organizations created to harness their energy and expertise.

The problem is neither new nor unfamiliar. Today, there is an almost universal recognition that the vast majority of the world's largest and most powerful organizations have lost much of the entrepreneurial spark and individual initiative that made them successful in the first place. Buried by the meddling interference of bureaucratic staff groups, isolated from vital resources by a fragmented organizational structure, and distracted from the outside world by a tangle of internal systems and procedures, thousands of frontline managers in large companies worldwide have neither the incentives nor the motivation to seek out emerging opportunities or pursue creative new ideas. Rather then working to realize that which would otherwise not happen, these managers are often toiling long and hard simply to bring about the inevitable.

Having recognized the problem, however, most companies have spent the last decade trying to sidestep rather than solve it. Acknowledging that their bureaucratic structure was short-circuiting the entrepreneurial spark plugs in their organizations, companies like Digital Equipment Corporation and General Motors created off-line "skunk works" to provide a more supportive environment away from the distracting demands and exasperating interference of a bureaucratic hierarchy, in which new ideas could be born and nurtured. Others like IBM and Kodak spun off newly established projects as independent units and deliberately decoupled them from the normal pressures and structures of the mainstream organization. And a few like Polaroid and Caterpillar even created new venture units aimed at replicating the greenhouse effect of a venture capital company by soliciting and funding new initiatives outside the normal capital budgeting system.

While each of these new approaches claimed some success in

triggering entrepreneurial activity, in very few cases have they provided the long-term, broad-based solutions companies were seeking. The problem is that all of these prescriptions are based on a common principle—rather than attempting to fix the bureaucratic structures and elaborate systems that have stifled entrepreneurial initiative, the companies chose to bypass them. As a result, while a company like IBM was able to develop and launch its personal computer in a spun-off company, this move did little to trigger individual initiative and creativity inside the main corporate structure. Indeed, it could not even protect the entrepreneurial spark in the personal computer group once it was reintegrated into the parent organization.

This was the legacy of the "organization man." The rational, authority-based corporate model that the professional management doctrine had framed worked only by making individuals "as predictable and controllable as the capital assets for which they were responsible," to recall Harold Geneen's words. Once the system had selected for and reinforced conformity and obedience, no amount of exhortation from top management about individual initiative or encouragement of personal risk-taking could change the dominant pattern.

Within this vast desert of organizational rationality and conformity, however, there existed a few isolated corporate oases where individuals routinely initiated innovative projects, took risks, and challenged the status quo. Perhaps the largest and most well-known of these was 3M, the company that had become the benchmark standard for those who rejected the notion that corporate entrepreneurship was an oxymoron.

INSTITUTIONALIZED ENTREPRENEURSHIP AT 3M

To understand how 3M has been able to incorporate entrepreneurial initiative into its organizational bloodstream, over the long haul, it is instructive to trace its development as it is parallel with—and in contrast to—the Norton Company, its arch-rival in the abrasives business. Roughly equal in size in the immediate postwar period, Norton was the more well-established company and 3M was the underdog, albeit an emerging challenger. By the mid-1950s, however, 3M had grown to twice the size of Norton; by the mid-1960s, it was four times larger; by the mid-1970s, it had six times the sales; and by the mid-1980s, it was outselling Norton eight to one. By the mid-1990s, while 3M was racking up an enviable record as a perennial entrant on *Fortune*'s list of the most admired companies in the United States, Norton had been swallowed up by the French industrial giant St. Gobain.

Throughout their respective histories, Norton and 3M have epitomized two very different approaches to managing large, diversified companies. Norton became an early and strong convert to the emerging doctrine of systems-based "professional management" soon after that approach was popularized in the 1920s. Early on, the company adopted the fashionable divisionalized structure, became an innovator in financial control systems, and pioneered the development of leading-edge strategic-planning systems such as the PIMS regression model and the growth-share matrix popularized by the Boston Consulting Group. Supported by expert staff-driven analysis, the top management of Norton consistently pursued a strategy of growth through acquisitions while simultaneously driving for profitability by monitoring the performance of its strategic business units against their defined portfolio roles.

In contrast, 3M's approach appeared far less sophisticated, particularly in an era where such tools and techniques were at the cutting edge of management practice. Indeed, there have been few companies whose start-up years were less promising than that of Minnesota Mining and Manufacturing Company. Struggling to generate a profit from its early mining operations, the company entered the sandpaper business, but its performance as a manufacturer was as unimpressive as its experience as a miner—at least until a young bookkeeper named William L. McKnight took charge. Rising to the top of the company in the 1920s, McKnight developed a management philosophy that was in many ways at odds with the leading-edge practices and principles emerging at the time. Yet as his beliefs became embodied in 3M's organizational norms and procedures, the company grew to be one of the most successful companies of modern times, and certainly one of the most consistently innovative.

By his own account, McKnight's notions about organization and management were profoundly affected by two individuals who were at the center of two key events in the development of the struggling young company. The first was Francis Okie, a brilliant but eccentric inventor whom McKnight hired as 3M's first dedicated product developer. Okie's first innovation was a waterproof sandpaper that found immediate market acceptance, not only confirming the value of research and experimentation but also establishing product differentiation as key to 3M's commercial success.

The second key event occurred in 1923 when Richard Drew, one of Okie's young laboratory technicians, was visiting an automobile plant on a routine service call. Drew observed that the workers were having difficulty with the new-style two-tone paint finishes, and the young technician was convinced he could develop a solution. Working with the adhesive compounds and coating processes that were at the core of 3M's basic technological expertise, Drew developed a product that was to become known as masking tape. The key lesson for McKnight was that by aggressively seeking out technological solutions to customers' needs the company could create valuable innovations, even on the basis of extremely limited core competencies.

Yet these two critical incidents in 3M's development had an even more profound impact on the company's management philosophy. The string of breakthroughs generated by the eccentric Okie (a man who, for years, sandpapered his face rather then shaved it) gave management not only a tolerance for unconventionality but a genuine appreciation for its genius. When coupled with Drew's outstanding success, which he later pursued in the development of a product launched in 1930 as "Scotch" brand cellulose tape, the events of the 1920s imprinted on McKnight and his management team an unshakable belief in the ability of the average individual.

As a consequence, while Norton was developing elaborate structures and sophisticated systems to help top management plan strategic objectives, allocate scarce resources, set operating targets, and control ongoing activities, McKnight was talking to his managers about their vital role in creating "an organizational climate that stimulates ordinary people to produce extraordinary performances." It was a management philosophy that focused more on recognizing the potential of each individual employee than on harnessing the power of the new structures and systems.

Over the years, the differences between the two companies' management styles became increasingly clear. In Norton, managers had tightly defined job descriptions; in 3M the 15 percent rule allowed anyone to spend up to one-seventh of his or her time pursuing personal "bootleg projects" that might be of potential value for the company. In Norton, elaborate planning processes and control systems guided resource allocation decisions; in 3M a philosophy of "make a little, sell a little" reflected management's belief that the market was usually a better judge of business potential than the management hierarchy, and that wherever possible, those with good ideas should be funded incrementally to allow them to prove themselves. And in Norton the divisional organization was just that—an organization divided from the top down in a symmetrical, logical hierarchy of tasks; in 3M, the organization was built from the bottom up on a "grow and divide" principle that allowed divisions to fund individual idea champions whose successful project teams became departments and eventually divisions, which then funded their own projects, in a continuing cycle of self-regeneration.

While 3M became a hotbed of sometimes frantic activity and disorganized experimentation, Norton developed as a more rational, logical, and neatly organized company, routinely acquiring companies to build the diversity it was never quite able to generate internally. By the 1990s, the entrepreneurial initiative of generations of "ordinary people" in 3M had created a portfolio of over 100 core technologies that had been leveraged into 60,000 products managed in 3,900 profit centers clustered under 47 product divisions. Yet despite its size and the maturity of many of its businesses, this seasoned industrial company, born in the sandpaper business, continues to grow through individual initiative that allows 3M to generate more than 30 percent of its sales from products introduced within the previous four years.

What explains the enduring success of 3M compared to the more typical struggle of its one-time nemesis, Norton? And how might others emulate 3M's powerful engine of institutionalized entrepreneurship? These are questions that benchmarking teams from scores of companies have tried to answer as they make their pilgrimage to Minneapolis–Saint Paul. Most come to a simple but immensely powerful conclusion: The secret lies not so much in the structures, programs, or incentives that frame 3M's ongoing activities, which do not greatly differ from the structural characteristics of other companies that have failed to create 3M's powerful and durable entrepreneurial engine. The difference is as simple as it is potent. At the foundation of everything in 3M is a deep, genuine, and unshakable belief in the ability of the individual. Surrounding it are a series of organizational policies and management practices that build on and leverage that basic belief.

Although the 3M culture is distinctive, its generic qualities are not unique. As we probed the operating environments of several other entrepreneurial companies, we found that they too had been able to develop the kind of ongoing frontline initiative that 3M had embedded over half a century. Drawing on the experiences of 3M and Danish-based cleaning services company ISS, and reviewing some of the lessons from the ABB story, we can highlight three common characteristics of these institutionalized entrepreneurial practices:

- Inspiring individual initiative requires that individuals feel a sense of ownership in what they do; this is achieved in smaller organizational units more easily than in large ones. Presented in this way, the point borders on the self-evident; yet, like Norton, most companies have created organizations that achieve precisely the opposite effect.

- To align frontline initiatives with the company's overall direction and to prevent distributed entrepreneurship from degenerating into chaos, the sense of ownership needs to be complemented by a strong sense of self-discipline. Self-discipline is the performance standard that comes from within each individual. Unlike control, it is not imposed from above.

- Management needs to reflect its respect for the individual in a supportive culture that is open to questioning from below and tolerant of failure. Only in such an environment are individuals really empowered, since they have the freedom to take the risks required in changing the status quo.

CREATING A SENSE OF OWNERSHIP

Ironically, in a careful review of the history of the modern corporation, one can observe that the gradual erosion of strong feelings of identity, belongingness, and even ownership that once existed between companies and their employees coincided with the growing acceptance of the power of delegating responsibilities. But, under the management doctrine that accompanied the arrival of the divisional organization model, the corollary of delegation was control. And it was the need for top management to maintain a tight check on the delegated responsibilities that led to the creation of new layers of hierarchy and

the more sophisticated systems that eventually choked entrepreneurship.

In its day, the new structure and systems-driven management
model was a powerful and effective innovation. But this approach
also had side effects that became increasingly debilitating over time.
When the information-sharing and decision-making processes in a
company are fundamentally changed, the relationship between
those on the front lines and managers further up the hierarchy
becomes strained. The frontline managers' frustration began when
they saw their creative thinking and careful analysis regimented by
standardized reporting formats and homogenized by a consolida-
tion process that abstracted and agglomerated information as it was
rolled up for top management review. This underlying frustration
turned into disengagement as the originating champions felt cut out
of discussions and review processes by the experts acting as filters,
evaluators, and interpreters of their ideas. Finally, disengagement
grew into dissatisfaction and full-blown cynicism when they saw
their richly developed plans and proposals reduced to a single number return in investment estimates or a floating bubble on a chart in
the company's strategic portfolio.

At this stage many frontline managers recognized that those at
the top were making decisions in isolation, on the basis of
abstract data that bore little relation to the ideas, dreams, and
passions of those actually running the operations. This caused
some to retreat into a mode of obedient conformity—they gave
up the dream and passively began filling in the numbers required
by the system's format. Others assumed a more defiant, subver-
sive attitude, seeking ways to "beat the system" by distorting
data, padding proposals, or making side deals. In either case, the
relationship between the individual and the corporation became
increasingly pathological and mutually destructive.

To counteract this disenfranchisement, ABB and thousands of
other companies like it are trying to reenergize individuals like
Don Jans who have become isolated, disengaged, and alienated
within an impersonal and unresponsive corporate entity. As they

do so, some have begun to discover the truths 3M has understood and practiced for decades. They have begun to focus again on people, recognizing that the first step toward reengaging individuals is to give them a sense of ownership. This has, of course, required that they rethink the organizational philosophy that made "the division" the basic building block of the modern corporation and replace it with a much smaller unit to which individuals can feel a greater sense of belonging and on which they can make a visible impact.

Creating Small Performance Units

Rather than being divided from the top down, the 3M organization grew from the bottom up, with individual initiatives giving birth to project teams that, if successful, developed into departments that in turn had the opportunity to grow into divisions. The motivation of ABB in breaking its $30 billion mass into 1,200 separate operating companies was to replicate the sense of personal identification and individual ownership that has provided the fuel for 3M's innovation engine. And it is the force driving hundreds of other companies to pull out layers of their hierarchy and rebuild the organization around small frontline units.

Yet as Norton and many other companies discovered, simply delayering, destaffing, and restructuring the organization is not enough to regenerate individual initiative. After all, the problem of frontline demotivation had been recognized for decades, and several previous attempts to deal with it by creating smaller units had failed to solve it. The proliferation of profit centers in the 1970s, for example, or the creation of strategic business units in the 1980s had brought about several important benefits, but the reengagement of individual employees and the rebirth of corporate entrepreneurship were not among them.

Like these earlier reconfigurations, which were driven primarily by administrative needs—profit centers to create more effective operating budgets and SBUs to generate better strategic plans—many of the current round of restructuring initiatives appear to be aimed primarily at making traditional structures and systems work more efficiently. In the transformation to the individualized corporation, however, an entirely different organizational logic is at work. It reverses the traditional perspective of allowing top-level executives to exercise their authority and maintain their control, and instead engages the energies and supports the initiatives of those on the front lines.

The corporate structure and management philosophy developed by Paul Andreassen, longtime CEO of Copenhagen-based International Service Systems (ISS), provides an excellent illustration of such an organization model. Andreassen has built a highly successful $2 billion company with operations spanning seventeen countries on three continents. In the hardscrabble commercial cleaning business traditionally dominated by fly-by-night cutthroat competitors and well-entrenched family businesses, this is an even more remarkable accomplishment. At the heart of ISS's success is an organization composed of small independent units to which employees feel an intense loyalty and where entrepreneurial initiative is allowed to flourish.

Beginning in the 1970s, Andreassen recognized that if he broke the company's monolithic structure, he could allow his frontline managers the opportunity to expand beyond ISS's traditional office-cleaning business into related areas such as hospitals, schools, and supermarkets. To communicate his vision, he created a new service-driven philosophy based on what he called "The Magic Formula" whose primary purpose was to help the organization create new opportunities by "nurturing the front line." Where once the company had operated through single national subsidiaries, each reporting through to three corporate divisions, Andreassen decided to break these powerful national entities into up to half a dozen small companies per country, focusing each on creating a distinct business built around a specific market segment. Despite the added cost of operating these small units, Andreassen believed that it would be more than repaid if it helped individual managers, supervisors, and even operating-level cleaners to develop the sense of identity and ownership necessary to grow the business.

His assumptions proved to be well founded as the company entered a phase of unprecedented growth driven by the energies of those closest to the customer. Employees in the supermarket-cleaning companies, for example, developed specialized capabilities for cleaning the meat section, freezers, and air-filtering systems, for a premium over the price of the standard cleaning service. From there, individual entrepreneurial activities expanded the business into labeling, uniform cleaning, and even the repair of shopping carts. Eventually whole new lines of business developed from these initiatives: a Dutch group opened up a major high-margin opportunity in the cleaning abattoirs, a U.K. company expanded into airport contracts, and a German team created a new business around the rubble disposal needs that emerged as the former East Germany opened to the West.

The point is not that companies like ISS, ABB, or 3M do not have large aggregated organizational units; they do. In fact, to the casual observer the organization charts of each of these companies would look every bit as hierarchical and bureaucratic as Westinghouse's or Norton's formal structure. The difference lies in how such units are perceived and therefore how they are managed. In 3M, for example, almost all of the company's fifty-odd divisions have developed around a particularly successful innovation, typically sponsored by a project team, then nurtured in a frontline department. In an organization born out of frontline initiative, the role of division vice-presidents and their relationship with the project teams and departments is fundamentally different from that of a division-level executive in charge of an administrative unit that has been superimposed on a portfolio of products or businesses arbitrarily deemed to be operationally interdependent or strategically synergistic.

In the end, it is all reflected in the managers' underlying assumptions about human motivation. In Norton's profit centers and Westinghouse's SBUs, corporate management designed measurement and reward systems in the belief that it was their role to try to motivate frontline employees; in 3M's project teams and ISS's operating companies, top management believed that it was their responsibility to create the working environment that would stimulate and support each individual to become self-motivated.

Radically Decentralized Resources and Responsibilities

The difference in beliefs about what motivates human behavior is also the other key reason why companies like 3M, ABB, and ISS were able to create highly entrepreneurial frontline units where Norton and Westinghouse were not. In the latter companies, top management clung to the conviction that it was their job to maintain direct control over the key resources in order to exercise their core responsibility as strategic architects. In contrast, the organizations that had been able to develop as truly individualized corporations recognized that those closest to the customers or most knowledgeable about the technology were usually far better placed to respond to fast-changing environmental demands or market opportunities.

This recognition, when coupled with an underlying belief in the individual, led to a radical decentralization of resources and responsibilities. Only when accompanied by such a major transfer of people, power, and strategic assets was traditional delegation transformed into legitimate empowerment. And only true empowerment could ensure that the energy of the small performance units was converted into added value rather than being dissipated through fragmentation.

Nowhere was the symbolism and the reality of empowerment more clearly demonstrated than at ABB, where Percy Barnevik underlined his commitment to reallocating power from headquarters management to frontline companies by slashing the size of the corporate-level staff of this $30 billion global company from over 2,000 to just 150 people. This led him to formulate what became known as "the 90 percent rule," a staffing guideline that was applied to the company's divisional- and regional-level activities, as well as to the headquarters operations of any acquired company. Under Barnevik's rule of thumb, 90 percent of any headquarters staff needed to move out of their traditional roles, either by transferring to the frontline companies where they could contribute directly to the operations or by spinning off outsourced companies, providing value-added services at competitive market rates, or by leaving the company altogether.

Not only did this radical reconfiguration of all headquarters functions generate huge savings for the company, it also forced a rapid redefinition of the relationship between senior-level managers at the division and corporate levels and the frontline leaders of operating companies and profit centers. Business heads and sector executives with staffs of only two or three were simply unable to intervene in the activities of the scores of independent businesses under them. Equally important, with the bulk of the company's resources under the operating companies' control—95 percent of the people, 90 percent of the R&D budget, and the financial resources derived from their ability to borrow and retain earnings—frontline managers had the freedom to make many of the decisions that once would have rattled endlessly up the hierarchy.

The strong philosophy of radical decentralization of resources and delegation of responsibilities is equally evident at ISS, where Paul Andreassen has long preached his gospel of faith in self-sufficient, free-standing businesses close to the customers, then supporting that belief by building "Chinese walls" to keep senior-level executives from interfering in their operations. And it is also reflected in the way in which 3M has built its valuable research and development resources and capabilities in one hundred or more specialized laboratories spread throughout the company. Located close to the frontline project teams that drive entrepreneurial activity, these labs continue the close-to-the-market innovative traditions established by pioneers like Okie and Drew.

Such decentralization of resources does not imply that front-line managers have complete autonomy in deciding on the use of resources, however. In ABB, ISS, and 3M, management has installed mechanisms to ensure that appropriate review and approval occurs. In ISS, for example, each national subsidiary is structured as a separate legal entity with its own board of direc-tors drawn from company executives familiar with the opera-tions and outside experts with valuable specialized knowledge or experience. Meeting on a quarterly basis, this board becomes the main forum in which local business heads can test ideas, solicit advice, and obtain approval on their strategic plans, operating budgets, and capital expenditures. ABB has created a remarkably similar local board structure for its 1,200 frontline companies. And 3M's "make a little, sell a little" philosophy encourages experimentation by reviewing project proposals early and often, then making incremental investment decisions biased in favor of the project champion's beliefs and commitments over purely rational analysis.

None of these companies has abdicated its authority over its frontline operators nor lost control over the assets and resources it has placed under their responsibility. Yet by moving the review and approval process deep into the organization, each company has been able to short-circuit the cumbersome and unresponsive resource allocation process that frustrates and demotivates so many frontline managers. In short, they have built processes that reflect their belief in the people working on the front lines and in doing so have strengthened the frontline managers' sense of ownership.

DEVELOPING SELF-DISCIPLINE

For those who have never worked in a radically decentralized organization where the people closest to the customer or the technology are given the resources and the responsibility to act, one question inevitably arises: How do you prevent such an organization from losing its sense of direction and degenerating into chaos? It is a question we have encountered dozens of times as skeptical managers question whether the approach—and the impressive achievements—of ABB, 3M, or ISS could be replicated in their own companies. Yet to those who have discovered the liberation of working in small units, the more legitimate question is: How can you create a sense of individual initiative in an organization where relationships are defined by hierarchical control?

The companies we have described as Individualized Corporations are far from organizational anarchies. ABB did not grow to dominate the highly competitive power equipment industry through the random activities of its 1,200 companies; ISS could not survive in its low-margin cleaning business without highly focused objectives and demanding performance standards; and 3M would be unable to maintain its ability to create hundreds of successful new businesses each year if its product development initiatives operated in unbounded chaos. Yet focus, direction, and performance in these companies are achieved not by retaining tight control over the strategic plans and operating budgets of the entrepreneurial entities but by embedding a sense of discipline in their ongoing routines and the everyday behaviors of individuals throughout their respective organizations.

While the nature of objective setting guarantees that there will always be some give-and-take negotiations between individuals and their superiors, the work environment in a company with discipline built into its operations minimizes the negative elements of the cops-and-robbers game playing that characterizes traditional planning and budget-setting processes. In companies like ABB and 3M, the organizational context of discipline is very different from the culture of control and compliance that permeated Westinghouse and Norton. In an environment where people enjoy more freedom, they go beyond the need to follow directives and conform to policies; in highly disciplined organizations they take responsibility for their own actions. It is a characteristic that is immediately apparent in the little things—people return phone calls promptly, they come to meetings on time, and most of all, they deliver on their promises.

Developing an organization where discipline is the norm takes considerable time and effort. But once created, it allows management to reduce the imposed controls and, in doing so, to release the energies of those working in middle-level and frontline positions. The most powerful means of building discipline into an organization are to establish clear performance standards, democratize information, and develop a context of continuous challenge based on internal peer comparisons.

Clear Standards and Expectations

Empowerment is not abandonment. One of the gravest errors committed by managers who saw the need to transfer assets, resources, and responsibilities to those below them was that they did so with few behavioral guidelines, under poorly defined outcome expectations, and with unclear standards of performance. Little wonder, then, that so many such experiments have ended in frustration on one side and disappointment on the other.

As managers loosen the tight reins of day-to-day control, they must replace it with a clear set of performance standards and behavioral expectations. More grounded and measurable than the vague objectives of a vision statement yet more broadly framed and durable than the line items of the annual budget, these standards and expectations are the criteria that set the height of the bar and define the conditions for autonomy.

Operating in the mature and cutthroat commercial cleaning business with its razor-thin margins and limited growth potential, Andreassen had established profitability and growth targets for his autonomous unit managers in ISS. As long as they met the

mandated minimum 5 percent profit before tax and the 12 percent growth target, these individuals could operate with almost
complete freedom behind the carefully maintained internal
"Chinese walls."

Over several decades, 3M had developed an even more elaborate set of performance standards to delineate expectations for its
businesses. Each of the fifty-odd divisions was expected to contribute directly to the corporate goals of an inflation-adjusted
growth in sales and earnings of 10 percent, pretax profit margins
of 20 percent, and a return on shareholders' equity of 25 percent.
But the company's best-known standard—and the one that reinforced the innovative initiative of its frontline units—was the
long-established requirement that 25 percent of sales come from
products introduced within the last five years. In 1992, 3M CEO
"Desi" DeSimone increased that goal to 30 percent of sales from
products four years old or less, effectively raising the new product
introduction bar by 50 percent.

The company was very clear that these standards were applied
uniformly across all of its fifty-odd major businesses—from
industrial abrasives to surgical products. Said one ex-CEO, "We
recognize some of our businesses as established, but none as
mature, and exempt none of them—not even the oldest—from
striving to meet our standards for growth and profitability."

Not all standards came in the form of performance objectives.
To earn the right to operate with relative autonomy, frontline
managers in the most effective decentralized companies operated
under a richly detailed behavioral contract with their corporate
leadership. In ABB, it took the form of a carefully detailed fifty-
five-page booklet known internally as the "policy bible." This
document presented in clear and precise terms the corporate values, organizational policies, and management practices expected
of the managers of the company's 1,300 legal entities. The document reflected CEO Percy Barnevik's belief that the more clearly
one communicates expectations, the more free an organization
can be.

Information Democratization

The support system that allows control-oriented top management to maintain its power is its privileged access to information
and data analysis. Indeed, in most companies, the information
systems have clearly been designed to serve the needs of one key
user group—the corporate-level executives. Weekly, monthly,
and quarterly reports containing the most detailed data from
every crevice and corner of the organization are then consolidated, analyzed, and refined for top management's review and
approval.

Meanwhile, managers in the frontline operating units feel frustrated and apprehensive. Why? Because they know that the primary purpose of the information they are providing is to monitor
and control their own activities. (In the words of one frustrated
manager, "Everything you report may be taken down as evidence
and used against you.") At the same time they know that the
huge amounts of time they spend gathering data and filling in
forms creates reports that are of little use in helping them run
their businesses.

One of the most powerful steps a company can take to convert
itself into an Individualized Corporation in which discipline is
built from within rather than being imposed from the top is to
challenge the basic design and utilization of the information systems. Beyond the obvious first step of eliminating dozens of
underused reporting formats, organizations in reshaping themselves as Individualized Corporations have undertaken a complete overhaul of their systems design to refocus on serving the
needs of the frontline managers.

ABB undertook such a revolutionary step with the design of its
well-known ABACUS system—an acronym which, interestingly,
stands for the ABB Accounting and Communication (*not* Control)
System. Developed under the assumption that "every line manager must learn to become his or her own controller," ABACUS
tracks thirty-two performance measures that can help frontline

managers monitor their business operations. Reports are released simultaneously to managers at all levels within the organization; top-level executives receive the same data in the same formats at the same time as those in the individual profit centers (albeit in a rolled-up presentation).

In ISS, a similar philosophy had been implemented in an even more extreme form. The company's Management Reporting System (referred to internally as MRS) was anything but a traditional upwardly focused control system. Its basic unit of analysis and reporting was neither the division, nor the national subsidiary, nor even the business unit, but the individual cleaning contract managed by a cleaning crew. The supervisors of these crews had been trained to read and interpret monthly MRS reports that detailed the budgeted versus actual costs by job. Soon, they engaged their cleaners in reviews of the data and discussions of how they might improve the profitability of their contract. As well as actively controlling costs, many of the cleaning teams began to talk to their customers about providing add-on services. And after reviewing the allocated overhead expenses of each successive layer of management, some even started asking their bosses to justify their value added on a particular cleaning contract.

A company's approach to information management reflects its basic assumptions about individual capabilities and human motivation. Freely distributed throughout the organization, it can become a powerful tool in helping individuals monitor and adapt their own performance; hoarded at the top of the organization, it is more likely to be used as a bludgeon to force the organization to conform to imposed objectives. 3M's assumptions about people's abilities are clearly reflected in the philosophy of 3M's William McKnight, who asked, "What is it about a business that we can decide at the top of the company that could not be decided just as well and much faster by those running the business if they had the same information?"

Peer Comparison Challenges

To be powerful, a performance standard must be legitimate. And, as many who have tried to impose challenging new objectives on their organizations have discovered, the basis of legitimacy is credibility. It is this understanding that has led so many companies to engage in a flurry of benchmarking activities, in the hope that employees will see Motorola's achievement of "Six Sigma" quality or 3M's ability to generate 30 percent of sales from new products as a confirmation that they can achieve the same.

Although such benchmarking comparisons have been both educational and inspirational for hundreds of companies, they have often run into more difficulty when used as a standard for objective setting. The problem is particularly acute in companies where the process of selecting the benchmarks and deciding on the appropriate new objectives and measures are driven by corporate staffs. Even where the line managers are involved in the process, it can be hard to convince the employees of a cement plant or tire company that a semiconductor company's Six Sigma quality or a diversified company's 30 percent new product hurdle is particularly relevant to their organization.

By contrast, the most disciplined entrepreneurial companies we studied used a subtle but much more potent variant of this approach to encourage frontline individuals, teams, and units to calibrate their performance against their most effective peers. Not only did this provide the legitimacy of comparing like with like, it also ensured the credibility that only comes from the ability of an individual to control the comparison.

One company that had built such a cross-unit peer-level calibration into its routine management processes was ABB. Leveraging the fact that the ABACUS reports were designed to provide business managers with the detailed performance data to control their own operations, senior-level managers running worldwide businesses handed off most of the monitoring and control back to the frontline units. Selecting the handful of key measures they felt were relevant to the business's development, these division-level executives created "performance league tables" that compared the performance of their frontline companies and distributed the results to them for review, comment, and action.

Simply compiling and publishing the comparative performance data was enough to trigger a flurry of corrective activity that proved far more effective than any staff-driven analysis or top management intervention. Not only were unit-level managers driven to move their companies up the league table—a motivation that became close to an obsession for those on the bottom rungs—but they were also able to identify directly comparable units at the top of the league tables that they could call on for advice and support.

A similar self-evaluating and self-correcting process was at work at ISS, where a peer-driven comparison system and cross-unit learning process minimized senior management intervention at every level of the organization. Supervisors of teams of cleaners used the MRS reports to identify how the margins on their cleaning contracts compared with those of their peers, triggering competitive but friendly rivalries and cross-contract practice comparisons.

Across similar ISS businesses, unit managers created strong internal networks that they exploited to raise performance to the level of the most effective company. For example, the manager of a neophyte train-cleaning business in the United Kingdom quickly identified the Danish operation as the best performer in his business and took a team to Denmark for two weeks to learn from it. Likewise, at the national subsidiary level, when the MRS system identified the Austrian company as the unit with highest profitability and sales growth, the managing director offered to share with his counterpart MDs details of the customer retention program that he believed to be the key factor in the Austrian success. Again, self-motivated learning proved much more powerful than top management intervention.

PROVIDING A SUPPORTIVE ENVIRONMENT

The changes we have described in the preceding pages—creating small decentralized units and defining the performance standards and information flows to support them—can go a long way in creating an organizational environment that fosters entrepreneurial activity. The problem is that after decades serving as loyal implementers in a classic hierarchy, most employees do not have the attitudes, knowledge, or skills to allow them to take advantage of the new freedom made possible by such changes to the structure and systems. To allow these individuals to become real frontline entrepreneurs, companies must create a nurturing and supportive environment that develops the skills and builds the confidence of those being asked to take on this new role.

Think about what is required to prepare an animal raised in captivity for release back to nature. It takes skill, patience, and a lot of time. In contrast to some of the popular mythology (or wishful thinking) of instant empowerment and overnight success, our observations led us to conclude that radical transfers of responsibility and power without adequate coaching and support is both naive and irresponsible. Indeed, it is the failure to provide adequate support that has led many to dismiss talk of empowerment as empty and even dishonest rhetoric, which typically masks a massive off-loading of additional duties on lower levels of the organization.

Goran Lindahl, Percy Barnevik's heir apparent as ABB's chief executive, has long conceived his role primarily as being teacher and coach to those reporting to him. By his own estimate he spends more than half of his time "developing engineers into managers and managers into leaders"—a very time-consuming process that requires him to carefully define and manage what he describes as "the framework" within which each individual should be allowed to operate freely. The challenge is gradually to loosen and eventually to remove the boundaries, controls, and restrictions, at which point the individual can be described as a true leader—in Lindahl's definition, someone who can take responsibility for setting and monitoring his or her own objectives and standards. "When we have developed all our managers into leaders," he says, "we will have a self-driven, self-renewing organization."

In all the companies where we saw frontline entrepreneurial initiative flourish, we saw management commit themselves to the development of two vital organizational attributes: an environment in which individuals could acquire the knowledge and skills to assume responsibility for self-management and control, and a culture that allowed them to build the self-confidence necessary for risk-taking. Perhaps more then anything else, it was the widespread willingness and ability of ordinary individuals at all levels of the organization to take and manage risks that separated the most entrepreneurial companies from the rest.

While their distinctive methods of achieving it varied, all of them shared several characteristics that seemed to be vital components of such an organization; they made one-on-one coaching a central part of the relationship between the manager and those for whom he or she was responsible; they created a legitimacy for people to break their traditional passive acceptance of authority by challenging and questioning decisions they felt were wrongheaded; and they established a tolerance for failure that provided an opportunity for risk-taking.

Management Coaching

In the radical decentralization of assets and resources that has accompanied the delayering and destaffing of so many companies in the past decade, the structural changes have generally been the easy part. While the accompanying layoffs often caused some trauma, management usually found it relatively easy to restructure their traditional hierarchies into flatter organizational configurations with more responsibility pushed down into small frontline units. The major difficulties typically followed later as they tried to redefine the roles of managers and realign the relationships between them.

For middle- and senior-level managers, in particular, the challenge of letting go of many of the controls that had previously defined their roles and provided them with their power and legitimacy was particularly difficult. As control over key resources and responsibility for vital activities moved to the frontline units, many of these managers felt they had become irrelevant. For them, the empowerment of some roles led to the evisceration of others.

In some companies, the perceived disenfranchisement of these middle- and senior-level managers became a major barrier to the overall transformation process. They became "the layer of clay" in the organization, blocking effective transfer of power down to the front lines and preventing the blossoming of new initiatives from below. These managers had not yet recognized that their control-oriented role needed to be restructured around a more supportive coaching-based relationship focused on helping the newly empowered frontline managers become the entrepreneurial engine they were expected to be.

ABB is very clear about its expectations of its middle- and senior-level managers. The company's "policy bible" defines their key role as being "to support and coach new managers." It also places great value on those who become "givers"—managers who have exceptional ability in attracting and developing talented people as candidates for positions in other parts of the company.

More than the formal policies, however, it is the espoused values of top management and the role model they provide that sets the standard of behavior in ABB. Lindahl's commitment to turn engineers into managers, and managers into leaders, sets the tone for his whole organization. So too does his way of dealing with performance issues. As he describes it, there are three questions he asks his managers when he detects a negative trend or an unexpected downturn in one of their businesses: "What is the cause of the problem? What are you doing to fix it? How can I help?" It is a very different approach from that which many of these individuals have experienced in an earlier time when corporate staff would be dispatched to analyze the problem and senior management would act on their analysis to intervene with corrective action.

Yet as most companies have clearly experienced—and as ABB would readily admit—changing the behavior of managers from one framed primarily around a control role to one built on a supportive coaching relationship is not always successful. Indeed, the biggest source of failure in the transformational change in most companies is not that those on the front lines are unable to rise to the challenge of becoming more entrepreneurial but that their managers fail in their ability to give them the freedom and support to do so.

In companies like 3M, where the coaching role has been built into the ongoing management processes over decades, the task is much simpler. Having been developed in such a system themselves, those who assume senior management positions understand that their key role is "to develop the people to develop the businesses," as one division vice-president put it. And doing so is much easier in an organization that has institutionalized a variety of policies and practices that support this key management role. For example, the company's "make a little, sell a little" philosophy not only defines an approach that allows 3M incrementally to fund new project proposals, it also creates a process that encourages managers to engage with a project team on a regular basis to review progress, discuss problems, offer advice, and set next-phase objectives. In other words, it injects coaching into the bloodstream of the relationship between the manager and his direct reports.

Openness to Challenge

Nothing is more tragic than a major corporate failure where most of those on the front lines could see it coming, while those at the top were either unaware of the problems or unsure of how to respond. It is a scenario that has been played out at scores of companies from General Motors to Philips, from Sumitomo to Apple Computer.

In such cases, the most common cause of failure is success. Management becomes so proud of its past achievements and so wedded to the strategic logic and organizational capabilities that made them that it loses the ability to reevaluate itself and revise its traditional approaches. To prevent the gradual ossification of corporate thinking into a process of cataloging and protecting "the company way," the most innovative companies ensure that their various priorities, policies, and practices are never so tightly defined that they become "off limits" to questioning, particularly from below. In establishing norms that not only allow but encourage open challenge among managers regardless of rank, they achieve two important advantages. Beyond the obvious benefit gained from subjecting outmoded assumptions to informed scrutiny, the process of challenge is enormously empowering to those at lower levels of the organization who historically felt their views were never heard. If creating small decentralized units is instrumental in creating a sense of ownership among these people, opening company policies and top-level decisions to question gives them a sense of membership.

Such an environment is well established in 3M, where, like most of the company's policies, it is rooted in respect for the individual. Instead of being indoctrinated through training sessions describing how to navigate through the procedures required to obtain formal approval on a project, the new 3M employee is likely to be regaled with stories about how legendary innovators challenged the system to get their ideas funded. They are likely to hear how Alvin Boese succeeded in perfecting the nonwoven fiber technology that later spawned products in nineteen different divisions in spite of three successive rejections of his proposals by management. Or how Philip Palmquist worked in his lab at night in defiance of orders to stop pursuing the reflective technology that subsequently gave birth to the whole range of Scotchlite products. Or how the project team dedicated to developing an insulating material continued their bootleg development despite management's insistence that this was not an appropriate business for 3M. The latter story is likely to be told by CEO "Desi" DeSimone—who then reveals that it was he who tried to stop the project that eventually culminated in the development of the highly successful Thinsulate brand of insulated outerwear.

DeSimone explained the delicate balance that management tried to maintain between ensuring disciplined implementation and maintaining an openness to challenge:

> Managers must retain a respect for ideas coming up from below. They have to ask, "What do you see that I am missing?" And they may have to close their eyes for a while, or leave the door open a crack when someone is absolutely insistent that their idea has value.

We saw exactly the same kind of openness to challenge from below in ISS where managers even felt comfortable questioning top management's major strategic decisions. When Andreassen decided that the company should get out of the security services business, for example, some of his best senior managers objected strongly. The head of the star Austrian subsidiary successfully argued that he should be able to retain the business that was operating profitably in his country. Reversing his blanket exit strategy, Andreassen agreed that it made sense to retain the business where it could operate successfully.

The willingness and ability of individuals to challenge embedded policies or to question top management decisions is made much easier when the routine interactions across organization levels are based on a coaching relationship rather than a relationship dominated by control. It is further leveraged as a source of confidence building when it is supported by an environment that exhibits tolerance for failure.

Tolerance for Failure

"Desi" DeSimone's oft-told yarns about 3M's most celebrated product champions who subverted the procedures, defied the decisions of top management, and generally battled the odds to defend their projects are really only half the story. The common thread that binds together all the legends of entrepreneurial innovation is that they all eventually succeeded. But what of the great many more stories of failure that surround those few glowing accounts that have been immortalized in 3M's corporate folklore? Interestingly enough, many of these are also woven into the company's oral history.

Most of these stories serve to illustrate the point that even those initiatives that seem to end in failure eventually give rise to a successful outcome. Sometimes, the entrepreneur succeeds in his next initiative; perhaps a new market application is found for a floundering product; or an undreamed-of application can often transform a failed technology into a success. The classic example of the latter, of course, is the failed experiment of a scientist who tried to develop a superstrong adhesive and ended up with a product with the opposite characteristics. Only later did another scientist by the name of Art Fry stumble on an application for a weak adhesive in a product that became known as Post-it notes. But as the old 3M adage goes, "You can only stumble when you are in motion."

So within 3M, top management saw a key part of its mission as legitimizing and even celebrating what was referred to within the company as "well-intentioned failure." It was another of the well-established values that had been instilled by the company's organizational architect and spiritual leader, William McKnight, whose philosophy was still cited within the company more than half a century after he articulated it:

> Mistakes will be made, but if a person is essentially right, the
> mistakes he makes are not as serious in the long run as the

mistakes management will make if it is dictatorial and undertakes to tell those under its authority exactly how they must
do their job. Management that is destructively critical when
mistakes are made kill initiative, and it is essential that we
have many people with initiative if we are to grow.

It should be clear, however, that creating an environment that
is tolerant of "well-intentioned failure" is not the same as creating a risk-free environment. While all the most entrepreneurial
companies we studied tolerated individual risk-taking, none of
them came close to eliminating the personal exposure that provides the energy to champion a new initiative. In ABB, Percy
Barnevik emphasized his 7–3 formula that encouraged managers
to make quick decisions that were right 70 percent of the time
rather than delay action in search of perfect information. But
while mistakes were often accepted of those initiating action,
others whose long-term track record showed that their judgment
did not justify management's faith were removed. Likewise at
ISS, where Paul Andreassen confirmed that he was careful never
to cut off a failure too early. But because the transparency of the
MRS reports allowed everyone to see when businesses were
struggling, eventually the risk-taking manager himself was forced
to face the reality. "There is a professional hazard involved in
being a leader," said Andreassen. "If he doesn't succeed, he won't
have the glory. And eventually he risks losing his stars."

RELEASING THE ENTREPRENEURIAL HOSTAGES

As the story of Don Jans demonstrated so clearly, inside every
corporate hierarchy there are entrepreneurial hostages striving to
break free. For over thirty years Jans toiled diligently on the front
lines of Westinghouse, living within the corporate policies and
procedures and loyally implementing the strategy and priorities
passed down from above. And in doing so, he was helping to har-
vest a business that top management had decided was mature.

Yet the turnaround that he achieved in the same business with the same management team in just three years was nothing short of miraculous. Clearly there was something different about the way ABB managed Jans and his team that energized them to achieve a new level of performance.

What we have witnessed and described is more than simple empowerment—at least as defined by the structural changes, job description redefinitions, and resource transfers that seem to dominate the thrust of most faddish empowerment programs we observed. In too many of the companies that have implemented such programs, the notion has remained an abstract one, viewed by the organization as empty rhetoric and often resulting in little more than additional cynicism at all levels.

Why is it that so few companies have been able to inspire the individual initiative and entrepreneurial activity that characterizes the operations of 3M, ISS, and ABB? We believe that the answer lies beyond empowerment programs. The missing ingredient is the belief in the individual that is at the heart of the individualized corporation.

True empowerment occurs only when it grows out of a reciprocal system of faith. Those deep in the organization must have faith in their company and its leadership, and senior management must have faith in the people in its organization. Working together, it is this mutually supportive system that allows the individual organization members to have faith in themselves, which in turn provides the engine for entrepreneurial initiative that drives the corporate engine.

This relationship of reciprocal faith is based less on intellectual understanding and rational agreement than on a strong emotional commitment. And after years of controlling hierarchical systems-driven companies where they have had little experience with spontaneous frontline initiative and entrepreneurship, most managers lack the basis on which to make that commitment. Yet, the responsibility for the first move in creating the spiral of mutual confidence clearly rests with top management. Only they can move beyond the banality of such homilies as "people are our most important assets" and initiate the series of changes we have described to the organizational infrastructure.

Like ABB, companies can unleash enormous potential by releasing the entrepreneurial hostages locked in their bureaucratic hierarchies. It all starts with a belief in people like Don Jans.

Creating and Leveraging Knowledge: From Individual Expertise to Organizational Learning

4

People are innately curious and, as social animals, are naturally motivated to interact and learn from one another. This is the second key assumption that shapes the philosophy of the Individualized Corporation, supplementing the fundamental belief in the power of individual initiative that provides the bedrock of entrepreneurial activity described in chapter 3. Over thousands of years, families, clans, and communities have evolved as teaching and learning groups, with individuals sharing information and synthesizing knowledge as a central part of their binding social interchange and as a key engine of their collective progress. Yet, somehow, modern corporations have been constructed in a way that constrains, impedes, and sometimes kills this natural instinct in people.

Focused on maximizing short-term static efficiency, most companies have been designed to extract as much value as possible from all their assets, including people. In that process, however, they have sacrificed the long-term dynamic efficiencies that come from continuously enhancing and upgrading the capabilities of individuals so as to enable them to create new value. The Individualized Corporation reverses the focus from value extraction to value creation by establishing continuous learning of individuals as a cornerstone of its organization—not just as a means to achieve its business objectives but as an end in itself.

Companies are more than collections of individuals, however, and inspiring individual initiative and learning is hardly a recipe for organizational effectiveness. For one thing, there is always the risk of the individuals moving on, taking with them the benefits of their ideas and expertise. There are few investment bankers today who do not live in the constant fear of losing their star trader to a higher offer, or worse still, seeing a whole team leave to join a competitor. Furthermore, because most large companies operate in industries in which economic benefit accrues to those who capture economies of scale or scope, individual capabilities and initiatives can rarely succeed in isolation, regardless of their quality and appropriateness.

These two factors have significant consequences for the concept of learning in a corporate setting. Beyond a commitment to developing people's initiative and expertise, the individualized corporation must be able to link dispersed initiatives and leverage distributed expertise, embedding the resulting relationships in a continuous process of organizational learning and action. What this means for companies and how they can develop this capability of collective learning is the topic of this chapter.

BEYOND STRATEGIC PLANNING TO ORGANIZATIONAL LEARNING

While investment banks, consulting firms, and science-based start-ups have long been acutely aware of the importance of individual expertise and organizational learning, in recent years, this

awareness has spread to companies in a variety of more traditional industries. Steel companies, historically reliant on the
economies of scale in their high-capacity integrated steel plants,
were shown the power of knowledge-based competition as Nucor
cut a swath through their markets on the strength of its rapid
learning cycle. And in the once stable energy business, the industry's traditional reliance on financial clout was shattered by
Enron's rapid expansion, based on aggressive entrepreneurship
and shared organizational learning.

One implication of this change has been a gradual fading of
corporate management's quarter-century-long love affair with
strategic planning. Two closely related forces have dimmed the
previous enchantment. On the one hand, the rapid pace of
change in the business environment has undermined the relevance of long-range plans that often were little more than projections of the past. And this, in turn, has forced managers to refocus their attention from a preoccupation with defining defensible
product-market positions to a newly awakened interest in how to
develop the organizational capability to sense and respond
rapidly and flexibly to change. As a consequence, managers
worldwide have begun to focus less on the task of forecasting and
planning for the future and more on the challenge of being
highly sensitive to emerging changes. Their broad objective is to
create an organization that is constantly experimenting with
appropriate responses, then is able to quickly diffuse the information and knowledge gained so it can be leveraged by the entire
organization. The age of strategic planning is fast evolving into
the era of organizational learning.

As in most long-term love affairs, the disengagement process
for most companies has been a protracted and difficult one. Like a
lingering box of old love letters, the deeply embedded structures,
systems, and processes that supported the strategic-planning
model have served as a constant reminder of the past, and as a
result, have become impediments to the development of a blossoming commitment to organizational learning.

Nowhere did we see this more clearly demonstrated than in Kentucky Fried Chicken (KFC), the U.S.-based fast-food chain whose aggressive internationalization in the 1970s made it a global leader in the franchised takeout restaurant business. While overseas managers had been given considerable freedom during the early expansion of operations abroad, in the early 1980s corporate headquarters staff was expanded from twenty to over a hundred people primarily to strengthen the strategic-planning systems, capital budgeting processes, and operational control mechanisms that linked the center to its overseas subsidiaries. Shin Ohkiwara, the entrepreneurial frontline president of KFC-Japan, reflected on the impact:

> The changes led to the imposition of many more ideas from the headquarters. Okay, we were willing to give them a try. But we kept thinking there were also lots of ways they could have learned from us. We felt our twelve-piece minibarrel could be a success elsewhere. And our small store layouts with their flexible kitchen design seemed ideal for U.S. shopping malls. We were even experimenting with chicken nuggets in 1981—well before McDonald's introduced them—but were told to stop.

The problem in KFC and in hundreds of companies like it was that the strategic-planning, capital-budgeting, and operating-control systems framed a very analytic management process designed to help top management make rational decisions about product-market positions and to measure performance in terms of financial criteria. For example, KFC's resource allocation systems provided corporate managers with the data to make logical site investment decisions, the strategic-planning process ensured consistency in product positioning across markets, and the operating-control mechanisms brought more uniform discipline to store-level activities worldwide. The difficulty was that these systems also made the management process more rigid and unadaptive, stomping out the chicken nugget experiments and the store design innovations that were later reinvented in the United States.

But we also saw several companies that had managed the transition to an organizational learning focus quite effectively. One such company was Skandia, a Swedish-based insurance company that lit a rocket under its traditionally conservative business by recognizing that it had to compete less on the defensibility of its market position and more on its organizational capability to adapt and learn quickly from the new companies it opened around the world. Propelling this aggressive international expansion was Assurance and Financial Services (AFS), a Skandia division that grew from a struggling start-up in 1986 to account for almost half the company's $7.5 billion revenues and 85 percent of its $100 million operating profits by 1995. Jan Carendi, president of AFS and architect of its growth, described the different mentality in his division:

> Creating bulletproof products and defensible strategic positions is yesterday's game. Today you need the ability and willpower to constantly develop and deploy new products that respond to changing customer needs. It requires an organization with the flexibility and competitive energy of a kid playing a video game rather than the analytical consistency of a grand master trying to hang on in a three-day chess match.

Both KFC and Skandia had developed the kind of entrepreneurial frontline operations we described in chapter 3. Yet only one of these companies was able to integrate those activities into an organization that could constantly develop and deploy new products. The other, meanwhile, became bogged down in battles between corporate headquarters and overseas subsidiaries that smothered the sparks of experimentation before they could ignite a fire of organizational learning.

What were the differences between these companies? And how did one develop such a valuable capability? Drawing mainly on the experiences of Skandia and McKinsey and Company, perhaps the world's best-known and most respected firm of management consultants, this chapter will address those questions.

LEVERAGING GLOBAL KNOWLEDGE AT MCKINSEY

Over the last decade and a half, McKinsey has converted a strong commitment to organizational learning into a powerful competitive tool not only to assert its intellectual leadership in the market but also to invigorate its organization internally. This is a capability that is visible not only in major firm offices like New York or London. It is a truly global asset that allows the most remote parts of the firm to capitalize on its vast intellectual capacity, as the following story of a typical McKinsey engagement illustrates.

For John Stuckey, a director in McKinsey's Sydney office, the invitation to bid for a study to develop a financial services growth strategy for one of Australia's most successful and respected companies was a source of enormous satisfaction. But it also presented him with a major challenge. As head of a medium-sized office in a small but diverse market, he had at his disposal relatively few consultants with financial industry expertise. Unfortunately, almost all of them were "conflicted out" of this engagement (to use the McKinsey terminology) because of work they had done for competing financial institutions in Australia.

Stuckey immediately began using his personal network to tap into McKinsey's worldwide resources for someone who could lead this first engagement for an important new client. After numerous phone calls and some personal lobbying, he identified Jeff Peters, a Boston-based senior engagement manager and veteran of twenty studies for financial institutions. The only problem: Peters had two prior commitments that would make him unavailable for the first six weeks of the Australian assignment.

Stuckey realized he would have to rely on McKinsey's knowledge network to support a relatively inexperienced local team through the initial stage of the study.

To get started, Stuckey and Ken Gibson, the local engagement director who would lead the project, identified a three-person team of available and nonconflicted associates. At the same time, they began assembling a group of specialists and experts who could act as consulting directors to the team. James Gorman, a personal financial services expert in New York, agreed to visit Sydney for a week and to be available for weekly conference calls; Majid Arab, an insurance industry specialist, committed to a two-week visit and a similar "on-call" availability; Andrew Doman, a London-based financial industry expert also signed on as a CD. In the Sydney office, Charles Conn, a leader in the firm's growth strategies practice, agreed to lend his expertise, as did Clem Doherty, a recognized thought leader on the impact of technology on strategy.

With guidance from Gibson, the team of three young associates began scanning McKinsey's various in-house directories for leads on new ideas, core documents, and designated experts. Their first source was the Firm Practice Information System (FPIS), a computerized database of client engagements that provided full information on initial proposals, final reports, and backup data resources developed on all client work done by the firm anywhere in the world. They then tapped into the Practice Development Network (PD Net), which contained over twelve thousand documents representing the processed knowledge and generalized insights developed by the firm's different practice areas. To identify internal McKinsey experts, they had access to the Knowledge Resources Directory, a small book that listed all firm experts and key document titles by practice area—in essence, a McKinsey Yellow Pages. In all, the team tracked down 179 relevant FPIS and PD documents and accessed the advice and experience of over sixty firm members worldwide. One team member described the experience:

> Ken was acting as engagement manager, but he was not
> really an expert in financial services, so we were even more
> reliant than usual on the internal network. . . . Being in a com-
> pletely different time zone had great advantages. If you hit a
> wall at the end of the day, you could drop messages in a
> dozen voice-mail boxes in Europe and the United States.
> Because the firm norm is that you respond to requests by
> colleagues, by morning you would have seven or eight new
> suggestions, data sources, or leads.

At the end of the first phase, the team convened a workshop designed to keep client management informed, involved, and committed to the emerging conclusions. From the forty-two ideas the study had generated, client participants focused on seven core beliefs and four viable options that provided the team's agenda for the next phase of the project. It was at this point that Peters was able to join the team. His wealth of experience allowed a rapid narrowing of the options and the emergence of some clear conclusions that the team presented to the board of the client company and that were fully accepted. The client's managing director described the outcome:

> We're a tough client, but I would rate their work as excellent.
> Their value added was in their access to knowledge, the intel-
> lectual rigor they bring, and their ability to build understanding
> and consensus among a diverse management group. . . . If
> things don't go ahead now, it's our own fault.

What is interesting in this story is that none of the three young associates nor the engagement director had any significant experience in the financial services industry. Yet, they were able to deliver outstanding value to a knowledgeable and demanding client on a highly specialized topic. In the course of our conversations not only with all the people connected with the project but also with others, including senior managers of the firm, it became clear that several characteristics of McKinsey lay at the heart of this success. These were similar to the characteristics we found in Skandia; in Intel, the world's leading semiconductor company; and in Andersen Consulting, the world's largest consulting firm:

- First, each of these companies invested very substantial resources in developing the expertise of their people. They went to great lengths to recruit the very best, and created structures and mechanisms that allowed their employees to continuously enhance, upgrade, and broaden their capabilities.

- Second, they established the tools, processes, and relationships necessary to support horizontal flows of information throughout their worldwide organizations to link and leverage individual knowledge and embed it in a collective process of shared learning.

- And finally, through supporting lateral sharing of knowledge and also as a product of that sharing, each of them had built a strong sense of trust, both among colleagues and between superiors and subordinates.

The overall effect of these three characteristics was that their organizations were built on a framework that appeared much more like an integrated network than a classic divisional hierarchy. It was this network structure that prevented the horizontal flows of information and knowledge from being swamped by the vertical ones, thereby serving as the anchor of their organizational learning capability.

DEVELOPING INDIVIDUAL EXPERTISE

Intel Corporation, the world's largest and most profitable semiconductor company, has a relatively young workforce. However, most of the technologies used in Intel's blockbuster Pentium chip did not exist when the scientists and engineers who developed it finished their graduate studies. Without recruiting the best brains in the business and without very large investments in training to keep them at the forefront of the rapidly evolving technologies, Intel simply could not survive in its business. Having the best people in the business and continuously upgrading their skills and expertise is not a "feel good" factor for Intel's management; it is a prerequisite for the company's survival.

In a service-based economy and an information age in which companies are increasingly competing on their ability to capture, develop, and apply scarce knowledge and expertise, recruiting and developing human assets are no longer a secondary function required to keep operations going. They are at the core of a company's competitiveness. Bill Gates understands this clearly and deems no activity more important than meeting with superior candidates to convince them they should join Microsoft. And T. J. Rogers, CEO of Cyprus Semiconductor, tells his managers that they should call him immediately, even is he is in a board meeting, if anyone offers their resignation. When he says that Cyprus employees are its most valuable assets, it is not empty rhetoric, as he confirms by his personal commitment to do whatever he can to hold on to them.

This is not just an issue for high-tech companies like Intel, Microsoft, and Cyprus, however. As we saw in chapter 3, managers in ISS, the industrial cleaning company, exhibit the same commitment. Recognizing that the quality of its people was the only thing that distinguished it from its competitors, every manager in the Danish company, from the managing director down, spent two days a month interviewing candidates to ensure that the company recruited the best people for the cleaning operator position. And through the company's Five Star training program, it developed its most capable recruits not only in cleaning techniques but also in quality control, customer relations, and even financial analysis. As a result, ISS had the most highly motivated and loyal labor force in the industry—not to mention an employee turnover rate less than half that of its competitors. More important, these motivated cleaning teams regularly became the source of superior cleaning performance and new business initiatives.

In the end, only through the superior knowledge, skills, and motivation of its employees can a company remain competitive. That is a truth that has become crystal clear to top management at Andersen Consulting. As Terry Neill, head of Andersen's worldwide change management practice describes it, Andersen's greatest challenge is "to stay ahead of the commoditization envelope." In the 1940s, Andersen's expertise lay in designing the complex manual accounting systems required to meet the expanded reporting requirements of the Securities and Exchange Commission (SEC) in the United States. When this capability became routine in the 1950s, Andersen moved past the many companies concentrating on this business to focus on the emerging need for payroll systems. But this work also became commonplace by the 1960s, leading Andersen into the development of computerized accounting and payroll systems. In the 1970s, the firm moved into systems integration and, by the late 1980s, had migrated to the even more complex task of business integration. It is clear to Andersen's leadership that, without constant upgrading of its people's skills, it could not have stayed ahead of the commoditization envelope to become the world's largest consulting firm.

Collectors of People

In the traditional corporate model, the company relies upon the wisdom and expertise of top management and the power of its established systems and processes to transfer and exploit that knowledge and experience in its day-to-day activities. The advantage of making such a model less dependent on the quality of anyone except the top managers is that, at least in theory, it can transcend any mediocrity in the middle and front lines of the organization. In contrast, the Individualized Corporation is based on an assumption of high quality and competence in all its employees. As a result, its effectiveness becomes highly dependent on the realism of this assumption.

A middle-level manager at Norton underscored the impact of the old assumptions in his observation of how his company's historic focus on capital assets contrasted with its relatively casual management of its human assets. To get approval for a new $100,000 machine, he needed to go through an elaborate planning process and seek approval across three layers of management. But there was no such interest, guidance, or support from senior management when he hired a new $100,000 senior engineer. Yet, as this manager pointed out, the personnel decision had a far higher total investment cost over the life of the "asset." And while the "right" decision on the machine would give the company no advantage over its competitors, an excellent selection choice could have a major positive impact.

The first requirement for developing individual expertise is to make recruiting decisions strategic decisions. As at Norton, the top management of most companies pays far more attention to building the most efficient plant or developing the best distribution network than they do to getting the best people. In contrast, in companies that focus on knowledge-based competition, top management recognizes that it can obtain sustainable competitive advantage by routinely recruiting people who are a little bit more skilled, motivated, or intelligent than the pool their competitors attracts. In essence, they become passionate collectors of people.

In Skandia, Jan Carendi is always on the lookout for first-class people to attract to his organization. When he spots an exceptional person, he makes a job offer, even when there is no obvious job for him or her. Such was the case with Ann Christian Pehrsson, a capable executive he met at a conference. Impressed by her quality of thinking and despite her initial reluctance,

Carendi pursued her until he could convince her to leave her job and join AFS. With no specific vacancy readily available, he named Pehrsson AFS's director of business development, a one-person function based in Stockholm. Over the years, Pehrsson converted this role into that of an intelligence center on new market entry, consolidating Skandia's worldwide knowledge base and codifying it so it could be channeled and adapted to any new start-up operation worldwide.

Beyond such personal head-hunting, top management in the most effective learning organizations placed an enormous emphasis on establishing institutionalized systems for recruiting the best talent in their firms. McKinsey's recruitment system is legendary, with the top graduates from the best business schools around the world consistently ranking the firm as their first-choice employer. Behind this ranking lies not only the firm's generous remuneration and outstanding training opportunities but also a huge investment of money and management time to support its recruitment process. Each year, the firm screens fifty thousand résumés and undertakes tens of thousands of interviews in order to hire five hundred new associates. Taking candidates through up to a dozen interviews puts huge demands on the time of McKinsey's five hundred partners, but all of them treat it as their number one priority.

The Company as a University

In the knowledge revolution currently unfolding, the half-life of any employee's base of expertise is often a few years rather than a few decades. As our earlier reference to Intel suggested, even the best and brightest new recruits become wasted assets from the time they join the company unless their skills are constantly upgraded. Companies such as Intel understand this and take their developmental responsibility seriously. For example, Intel operates its own university, offering a plethora of courses to which employees can self-enroll; it supports participation in a large variety of external courses delivered by universities and consultants; and it offers a sabbatical scheme so that people at all levels of the company can take a few months to a year off to go back to school, either as a faculty member or as a student.

As the Intel example illustrates, this strategy is not about offering occasional seminars or creating ad hoc training programs on topics of current interest. To create an organization that is able to learn, a company must develop people who are hungry for knowledge. There is no better way to keep their appetites whetted than to continually stimulate them through an education and development agenda that is built into the organization's business infrastructure. General Electric's world-renowned Crotonville faculty and the much imitated Motorola University stand as symbols to the commitment that these companies have made to the ongoing development of their employees. And the outstanding success of these companies in developing unmatched management depth is a testimony to the power of that commitment.

Building employee development into the ongoing routine of corporate life usually contributes to learning in ways companies do not anticipate. When McKinsey decided to dramatically increase its commitment to the training of its associates in the late 1970s, it found that by forcing partners to take on the role of professors, they began to articulate and document knowledge that had long been tacit. Equally important was the impact on the participants who not only learned about new tools, models, and frameworks but also developed the contacts and relationships that became a vital part of the firm's ability to develop and diffuse knowledge rapidly around the world. Today the firm commits over 5 percent of its $1.8 billion annual revenue to the development of its four thousand professional staff, an investment of over $20,000 per consultant per year.

One of the great benefits of this investment in individuals' development is that in addition to adding to their value as human assets, such commitment builds strong bonds of personal loyalty to the organization. Andersen Consulting has a commitment to education every bit as strong as McKinsey's, and each new con-sultant can expect to spend almost a thousand hours in training during his or her first five years, much of it on the campus of the firm's own state-of-the-art facility in St. Charles, Illinois. In spite of this huge investment in its people, Andersen lays no propri-etary claim to the skills and knowledge it helps its people develop. It is simply seen as an expression of its commitment to make its consultants the most competent and professional in their field. But despite the promise of the recruitment brochure that "after training with Andersen, you could work for anyone, anywhere—or you could work for yourself," the organization's willingness to invest in them makes the vast majority of recruits want to stay.

The more a company sees itself as competing on the basis of its superior information, knowledge, and expertise, the more it will have to view itself as something other than just a portfolio of financial resources to be distributed in the most efficient manner. As will be explored in more detail in chapter 5, a company that has taken these concepts further than any other we studied is Kao, the Japanese consumer packaged-goods company. It sees itself not as a soap and detergent company but, above all, as an educational institution. And it is clear to all within it that the two most important responsibilities of any manager are to teach and to learn. It is a powerful concept for a company, and as we shall see, one that has allowed Kao to break away from its competitors and become one of Japan's most innovative and creative compa-nies.

New Developmental Career Paths

No matter how well they are designed and delivered, the best that training programs can do is prepare an individual for the real learning that initially must come in applying the acquired knowl-edge on the job. And yet very few companies explicitly recognize that the organization's work can and should be constructed not only to achieve the required product or service output but also to develop the productive resources—the employees. Indeed it is often the case that only after the work is reconfigured to develop those performing it that the desired product or service can be provided—a classic case of strategy following structure.

This was clearly true when McKinsey faced the biggest challenge in its history in the mid-1970s. At that time, a group of highly focused competitors such as Boston Consulting Group began to make strong inroads into McKinsey's markets. On the basis of some simple but powerful tools like the experience curve and the growth-share matrix, BCG developed an approach it described as "thought leadership" to win clients and young recruits away from McKinsey, which in contrast founded its practice on a commitment to building client relationships. After several years of stalled growth and internal turmoil, McKinsey's partners recognized that the firm could no longer succeed simply by building strong relationships and assigning intelligent generalists to increasingly specialized problems. It would have to develop what they referred to as "T-shaped consultants"—individuals who supplemented their broad generalist perspective with an in-depth "spike" of specific industry or functional expertise.

While retaining the offices that had been home to the client-focused generalists—an important asset—the firm overlaid that basic foundation with an industry sector organization and a functional competence structure. In both cases, the overlays were designed to provide the superstructure that would support consultants as they developed new organizational relationships and career path opportunities. It provided them with opportunities outside their home offices that would help them develop their expertise spike in a way that they never could in a constrained geographic location. By the mid-1990s, nearly 20 percent of work was performed by consultants on short- or long-term transfer to another office, moves that were inevitably designed to develop or deploy individual consultants' specialized knowledge or expertise.

A similar motivation was driving Skandia's Carendi when he created the Skandia Future Center, a virtual organization initially built around five "future teams" designed to stretch, challenge, and develop the high-potential employees assigned to them. Each team was composed of five full-time members selected from across the organization and representing three populations—the "in power" generation, the "potential" generation, and "generation X," the twenty-somethings who represented the company's longer-term leadership pool. Far from being just a training exercise, this was a four-month, full-time developmental assignment for these team members. Each team was assigned an issue of great importance to Skandia's future—the impact of changing demographics on the insurance market, for example, or the future use of information technology in the company's organization and strategy. The process was not only enormously broadening for those involved, it was also energizing and motivating, providing them with new ideas and fresh commitment that they could take back to their jobs.

DEVELOPING HORIZONTAL INFORMATION FLOWS

As powerful a tool as the recruitment and development of superior individuals can be, in the end a company cannot gain advantage from accumulating islands of information and pockets of expertise that are of little value outside their own isolated area of responsibility. Only when it develops the ability to transfer, share, and leverage such fragmented knowledge and expertise will the company be able to exploit the benefits of organizational learning. For that, individual expertise in isolated units must be linked in a rich horizontal flow of information and knowledge that can routinely diffuse critical expertise and transfer best practices organizationwide.

As simple as it sounds, this has turned out to be a task that most companies find extremely difficult. The reason is that in traditional divisionalized hierarchies, information flows usually are framed by the vertical reporting relationships and the formal planning and control systems that support them. As we saw in chapter 3, for over three-quarters of a century, these systems have been refined and reinforced to become the informational packhorses for organizations, hauling plans, proposals, and requests up to senior-level and top management, and bringing down the objectives, allocations, and controls that define the behavior of those on the front lines.

The challenge facing companies trying to create an environment that would foster organizational learning was that they had to supplement, and in some cases supplant, these deeply embedded information flows with communications channels that were radically different. While traditional processes were, for the most part, framed by vertically driven formal systems and calculated in financial terms, cross-unit learning required information channels that were more often built on horizontally based personal relationships, through which knowledge and expertise were commonly transferred.

As we saw in chapter 3, 3M had developed such a strong horizontal information transfer process, particularly among its technical employees. Scattered across 3,900 profit centers in forty-seven divisions, these individuals were bound together by a strong belief that "products belong to divisions, but technology belongs to the company." Others like Kao and McKinsey had struggled for decades to create the same kind of free flow of knowledge across organizational boundaries, finally succeeding in making it the primary source of their competitive advantage. And still others like Skandia's AFS division and Intel had been built from scratch around a fundamentally different model of information transfer and communication processes. Common to all of these companies, and others we studied, were the creation of new channels of communication, the formalization and legitimization of horizontal linkages, and the development of new and different dimensions and metrics against which the organization measured and reported performance.

New Channels of Communication

In an era in which the computer has allowed managers to move vast amounts of data around their companies at the click of a mouse button, many have erroneously assumed that their embedded base of knowledge would follow. But knowledge is a notoriously "sticky" asset. Dammed off by proprietary protection on one side and a not-invented-here mentality on the other, it tends to accumulate in pools rather than trickle across the organization, causing new ideas to sprout and bloom in the once barren deserts of the frontline operations. The most basic task in creating horizontal information flows, therefore, is to create new channels of communication that encourage the rapid diffusion of strategic knowledge and expertise across the organization.

The simplest and most direct way to create such cross-unit linkages is for senior management to play an active personal role in leading the cross-fertilization process. Through his own initiative and by personal example, Jan Carendi became an extremely powerful force in the efforts to leverage the AFS division's growing knowledge and expertise across its worldwide federation of unit trust insurance companies. Indeed, he saw himself not as some decision maker sitting at the apex of his company but as a mobile resource and agent of change, available to provide advice, bring support, and connect diverse resources and activities. "I am," he said, "a civil servant to the organization."

To fulfill his linkage role, Carendi took on multiple functions at many levels throughout the company, always trying, in his words, "to infiltrate the organization." In addition to his role as CEO of the worldwide Assurance and Financial Services business, he served as a member of Skandia's corporate executive committee, as board chairman of AFS's three regional holding companies, as CEO of American Skandia, and even as president of American Skandia's investment company—roles that allowed him to penetrate five levels of the company's hierarchy and stretched him across all three geographic regions of its market reach. Wherever he went in the two hundred days a year he spent traveling, he preached the message: "Innovation does not mean starting out with a blank page. Good ideas come from other places and work previously done must be captured and reused." Through his own boundary-spanning role model, he became an active cross-pollinator of new ideas. More important, he represented a role model and catalyst, leading others to reach out beyond the parochial perspectives of their own job descriptions and embrace people and ideas from other parts of the organization.

Ingvar Kamprad, founder of IKEA—the company that revolutionized the traditional business of making and selling furniture—also recognized the power of personal networks and liked to transfer ideas and cross-fertilize best practices on what he described as "a mouth-to-ear basis." Reaching beyond the personal role that Carendi played, Kamprad supplemented his own prodigious personal communications efforts with individuals he described as "culture bearers." These were high-potential and experienced employees whom Kamprad prepared for their future roles by helping them to internalize the company's values through weeklong training sessions that he personally led. By the early 1990s, the company had assigned more than three hundred of these "ambassadors" to key positions worldwide, creating a dense personal network that could collect, interpret, and transmit information without the distortion that formal systems often introduce.

But cross-unit communication requires more than just informal ambassadors and personal networks to carry the volume and complexity of information that must be transferred among units. With the progress in information technologies, it has now become possible to support these informal horizontal linkages through a variety of tools and systems. After a major internal study, McKinsey recognized the need to supplement its largely informal processes for transferring knowledge between offices with a set of more formal activities. This is how the firm came to develop tools like the Firm Practice Information System (FPIS), the Practice Development Network (PD Net) and the Knowledge Resources Directory that the young team in Sydney found so useful.

Formalized Horizontal Linkages

As powerful as such cross-unit communications arrangements can be, most companies we looked at found that, in the crunch, the vertical information flows framed by the hierarchical reporting relationships would swamp the less well-established horizontal links. To give muscle to the newer channels and forums of exchange, many companies found they had to formalize the cross-unit relationships that gave them life.

At its simplest level, this involved creating teams and task forces drawn from different organizational entities and giving them responsibility for tasks that required cross-unit collaboration. Skandia employed this device frequently as a model to capture the knowledge and experience of multiple units and bring it to bear on the start-up of a new subsidiary. Living by his belief that "work previously done must be captured and reused," Carendi deliberately worked on developing a prototype for the organizational framework, administrative systems, and product design characteristics that could be transferred, adapted, and refined as the company continued its worldwide rollout. Teams from established subsidiaries became resources to the new start-ups, available to provide them with their updated models of the core prototype and to act as advisers and support to the new ventures.

McKinsey had taken an even stronger approach to formalizing its nonhierarchical relationships. It created the industry and functional specialization groups we described earlier and used them to reinforce the loose, informal linkages that existed among various industry and functional specialists, often isolated within their individual local office structures. It was in one of the original practice development teams that Tom Peters and Bob Waterman began to develop the ideas later published in the first management blockbuster, *In Search of Excellence*.

Supported by the information-transferring infrastructure of PPIS, PD Net and so on, these overlaid relationships gradually took on the same importance as the traditional geographic office connections. Partners were assigned practice leadership roles for each of the industry sectors and centers of competence; full-time practice coordinators were hired to monitor the quality of information flows and help consultants access relevant expertise wherever it existed; and new career paths were created to legitimize the development of practice-dedicated specialists alongside the firmwide generalists who had traditionally populated the partnership. In short, McKinsey formalized the development of cross-unit practice specialties that provided the framework for firmwide knowledge dissemination and learning.

New Metrics and Measurement

Despite the various formal and informal linkages to encourage more horizontal communication and more cross-unit integration, for many companies the elusive quest for organizational learning was often blocked by the parochial attitudes and behaviors induced by the vertically directed, financially driven measurement systems. By the early 1990s, managers were beginning to recognize that a laserlike focus on return-on-investment objectives and budgeted profit targets was often leading to pathological behaviors within the organization. Recognizing this problem, some managers began experimenting with "balanced scorecard" methods of measuring performance, while others introduced 360-degree evaluation processes as a way to break the tyranny of hierarchical appraisals. Such radical breaks with tradition did much to open up the horizontal information flows so vital to the creation of a knowledge network.

At the simplest level, change can be initiated simply by using existing measurement systems in a different way. For example,

BancOne built its highly successful interstate banking network on the foundation of acquired local banks. It then linked them together and had them learn from each other in a horizontal information exchange process it termed "share and compare." Central to the process was a standardized management information and control system that measured the performance of affiliate banks on scores of traditional measures such as loan quality, liquidity measures, productivity data, and earnings and expense ratio. What was unusual was that these reports were prepared not so much for top management control as for peer review. Dividing the ninety-eight banks in its network into three peer groups according to asset size, management distributed a two-page comparison report that allowed affiliate banks to compare their performance on key dimensions with those of similar-sized sister banks. Management then encouraged transfer of best practices by creating new communication channels (bank president council meetings, for example) and legitimizing the new horizontal relationships (such as formal mentor bank relationships). But the ability of these new arrangements to ensure the transfer of best practices effectively relied on the motivation of the receiving bank to reach out and ask for assistance; BancOne management believed that its "share and compare" measurement process provided that vital demand pull.

We observed an even more fundamental revolution in management metrics at Skandia. Here, Jan Carendi became convinced that the company's competitiveness depended on the value of its knowledge assets. Yet, like any company in the financial services industry, Skandia's accounting systems were denominated almost exclusively in financial terms. Worried that traditional systems did not even recognize, let alone try to measure, the knowledge flows and intellectual capital that he thought were vital to the company's future, Carendi hired Leif Edvinsson to become director of intellectual capital with the challenge of developing an alternative measurement system and new management processes that would focus on these vital issues.

Given free rein and strong support by his boss, Edvinsson created a rich conceptual model and language system that started managers talking about this unaccounted-for asset. But the real power came when he won Carendi's support to hire a controller for intellectual capital, allowing him to define measures and gather data that moved discussions from an academic exchange to a hard-nosed evaluation of performance. Using a device he called the Business Navigator, Edvinsson began tracking changes in performance on five key dimensions—the traditional financial focus (which he felt recorded yesterday's performance), the customer focus, the human focus, and the process focus (reflecting today's performance), and the renewal and development focus (that Edvinsson believed represented tomorrow's performance). These measures required managers to define and track their activities and achievements on dozens of nonfinancial dimensions such as the number of points of sale and the contract surrender ratio for the customer focus dimension, manager-to-employee ratio and training expense per employee for the human focus measure, and percent of premiums from new products and business development to total administrative expense ratio for the renewal focus indicator.

As Carendi confirmed, to focus managers on the vital task of building knowledge and transferring learning, he was trying to do nothing less than redefine the metrics against which the organization measured itself. As Edvinsson summarized, "We see the bottom line financial results as our top line. The real bottom line we want to focus on is development and renewal—the foundation for the future." Such a profound shift in strategic objective could never have occurred without an equally radical change in the way the organization measured and evaluated its performance.

CREATING A TRUST-BASED CULTURE

The hardest part of building a learning organization, however, is to create a culture in which individuals must share information or expertise that was once a major source of their power, accept responsibility for issues over which they have only limited control, and propose initiatives and take action in an environment in which the measures and metrics are unclear or in transition. Such behavior cannot thrive in an organization in which relationships are primarily contractual in nature and individuals and operating units are motivated to protect their self-interest. Instead, it requires more organic, familylike emotional bonding in which people rely on each other's judgment and depend on each other's commitments. In short, it requires a culture based on trust.

Trust was a clear element in the culture of all the companies we studied in which knowledge transfer and organizational learning were at the heart of their strategic capability. At 3M, for example, a strong trusting relationship between senior managers and those on the front lines provided the context for individual initiative, while a shared confidence among those who worked together across organizational boundaries framed the environment for interunit support. On the organizational trapeze, individuals will take the entrepreneurial leap only if they believe that there will be a strong and supportive pair of hands at the other end to catch them.

Most companies have not yet developed the kind of institutionalized trust that 3M has been able to foster over more than half a century of implementing a philosophy based on a belief in the individual, and in creating a climate designed to "stimulate ordinary people to produce extraordinary performances." Indeed, in the process of downsizing and restructuring their organizations, many companies have done more to destroy trust than reinforce it.

Nonetheless, in a number of companies we observed management working diligently to develop the kind of trust-based culture vital to the successful operation of an organization seeking to develop intensive cross-unit information flows. The most important factors we identified were a transparency and openness in organizational processes, a sense of fairness and equity in management decision making, and a shared set of core values that had been established within the organizations.

Transparency and Openness

Trust is most easily recognized in the transparency and openness of management processes that provide employees with a sense of involvement and participation in an organization. To a great extent, such a feeling emerges from the company's embedded norms and values, factors that can be greatly influenced by top management's role-modeled operating style. For example, the constant traveling that Jan Carendi undertook put him in regular contact with people throughout Skandia's worldwide operations. His natural, questioning, challenging style soon spread a message of openness in which honest disagreement became a powerful driver of learning. Like other mobile managers in the company, Carendi also became a cross-pollinator of information—ideas picked up in one part of the organization were dropped down in another. As much as anything, it was his inability to hide his views or hoard information that was at the foundation of the trust relationship he had developed so quickly with his management team.

Skandia also used its advanced information technology to create an open internal environment by linking its worldwide companies in a single transparent communications network. Its custom-designed global area network (GAN) connected individual subsidiaries' local area networks to create a worldwide infrastructure used to transmit electronic mail, act as a central repository for externally sourced information, provide electronic best-practices bulletin boards, share documents and files, and keep its traveling executives constantly in touch, wherever they were in the world. Furthermore, when issues required joint action, people anywhere in the organization could nominate candidates or volunteer themselves to work on the problem. The final composition of the team would be determined by polling nominees for their input on who could best serve the team. In this environment of broad involvement encouraged by organizational transparency, those at the top of the organization focused primarily on framing the issues, then opening up the process for resolution. Explained Edvinsson, director of intellectual capital, "We don't refer to Stockholm as the *head* office. The brain power is out in the field. If anything, the center acts as the *heart office*, maintaining the values of the group and helping to pump information—our lifeblood—around the organization."

But trust cannot be quickly imposed on an organization simply by holding regular meetings and creating open communication systems. It must also be built through the way people are selected and relationships developed. It is a concept that is well understood at McKinsey. During decades of emphasizing the "one firm" concept, McKinsey has built a culture in which mutual respect and shared trust characterize the partners' relationships with one another. To protect this valuable asset, the "spirit of partnership"—defined as openness, trustworthiness, and personal integrity—is an important qualifying criterion for election to the firm's management group. The result: a self-reinforcing environment that ensured alignment and mutual support on one hand, while on the other encouraged what was described as "the obligation to dissent" without destroying the underlying trust. In consequence, the firm has developed a robust model of self-governance in which partners debate vigorously vital issues such as the firm's desirable size and growth rate, partnership promotion criteria, and all other major strategic and policy decisions.

Fairness and Equity

Although transparency and openness are vital to allowing people to observe and take part in important organizational processes and management decisions, trust can be built and maintained only if the processes and decisions are recognized as inherently fair. Particularly in emotionally sensitive issues such as plant closings or layoffs, decades of investment in trust can be destroyed overnight if the decisions are viewed as insensitive, expedient, or politicized.

In the memory chip bloodbaths of the mid-1980s, when semiconductor manufacturers were forced to lay off thousands of workers as low-priced Japanese chips flooded the market, Intel built rather than destroyed employee trust by going to great lengths to find other solutions. Alternatives such as selling off 20 percent of the company to IBM to raise funds and implementing across-the-board pay cuts, not sparing management, to bring down costs were exercised before facilities were finally closed down. In Philips's semiconductor business, a 20 percent reduction in personnel was achieved without destroying trust by making the layoff decisions in collective meetings based on objective data using benchmarked performance rather than in "corridor deals" cut between people protecting their self-interests.

Like many professional service firms, McKinsey faces this issue continuously, because its entire personnel system is based on the principle of "up or out." With fewer than 20 percent of recruits being elected partners, the firm goes to extraordinary lengths to make its performance review process unquestionably fair and its separation process sensitively humane. The process begins from the time the new consultant joins the firm, receiving day-to-day informal feedback from almost anyone he or she works with. In each client engagement, formal development objectives are set for each consultant within the first week of the study, leading on to mid- and end-study progress reviews by the engagement manager. At least once a quarter, each consultant receives career feedback and guidance from his or her designated development leader, the partner assigned to each individual on the basis of common interests. Twice a year, the consultant receives a more formal review of performance and is given the opportunity to revise his development plans for the next six months. After five or six years of such feedback and development, the partners in an office decide if the individual is ready to be nominated for election as a principal, the firm's entry level into the partnership. Equally important, however, is the fact that with all the feedback the young associate most often has reached a similar conclusion about his or her readiness.

The Principal Candidate Evaluation Committee (PCEC), comprising twenty of the firm's most senior partners, assembles twice a year to evaluate candidates, each of whom has been screened and nominated by one of McKinsey's sixty-nine offices worldwide. Each PCEC member leads the evaluation of six to eight candidates from offices other than his or her own, spending approximately six weeks of full-time effort in the process. The subsequent report, recommendation, and vote by the full partnership ensures that the election process is as rigorous and fair as it can be. A similar process is conducted for election to director, the next level of the partnership.

Such is the perceived fairness of the McKinsey system that most of those who do not make it to partner maintain extremely strong relationships long after their departure from the firm. Indeed, many of its ex-consultants, after reaching top-level corporate positions, hire McKinsey to advise them. Ex-McKinseyites have become one of the firm's major sources of business.

Shared Organizational Values

During the past decade, many companies have worked hard to establish a vision statement that would generate excitement and commitment among employees. Yet when added to existing strategic objectives, budget targets, and program priorities, such grand statements of what the company wishes to achieve sometimes stir up more of a sense of cynicism than optimism—a feeling that this is just another means of driving the organization harder. In the words of one such disaffected manager, "It is fine to emphasize what we must shoot for, but we also need to know what we stand for."

It is a shared set of values—"what we stand for"—that creates the trust that is necessary to counterbalance the stretching effect of the vision's ambitious goals. Yet, identifying, communicating, and shaping a company's organizational values is often a far more difficult task than articulating a strategic vision since it relies less on analysis and logic and more on emotion and intuition. Although every well-established company operates on a set of beliefs and philosophies about what it regards as important, in most companies they remain implicit and in some they are deliberately repressed so as not to distract the organization from its focused business agenda or create an unnecessary obstacle to management actions.

Companies that have transcended this state of philosophical sterility to assert more boldly "what they stand for" typically attract and retain employees who identify more strongly with those values, thereby becoming more deeply committed to the organization that embodies them. "In the end," says Goran Lindahl, ABB's president, "people are loyal not to a particular boss or even to a company, but to a set of values they believe in and find satisfying." Shared values lead to a collective identity, a sense of unity and solidarity that facilitates trusting and sharing and thereby supports the horizontal flows that are so vital for organizational learning.

Although James O. McKinsey provided the vision that drove the firm bearing his name, it was Marvin Bower, McKinsey's managing partner from 1950 to 1967, who articulated its values and embedded them in the firm. Bower determined that this group of "efficiency experts," as they were often called, needed to become a firm of professionals, with standards of personal integrity, technical excellence, and professional ethics equal to those of an elite law partnership. Only then, he believed, would the firm be able to attract and develop associates of outstanding ability and clients of stature and importance.

At the heart of these values was the "one firm" principle that required all consultants to be recruited and advanced on a firmwide basis, all clients to be treated as McKinsey's responsibility, and all profits to be shared from a single firm pool. Bower believed strongly that only by operating in this way could McKinsey ensure that its professional standards, its commitment to clients, and its spirit of partnership would be maintained. Of course, such sentiments could have ended in nothing more than the themes of a few long-forgotten speeches had Bower and his successors not converted the fine words into operating practice. Each of the firm's leaders over the years has worked to ensure that every member of the firm understood that their membership in McKinsey meant a commitment to making a positive and lasting impact on their clients' performance *and* to helping build and maintain a great firm able to attract and retain exceptional people. Given the firm's rapid growth, for each generation of managing director the task of creating the free flow of people and ideas and the uniform standards of excellence implied in the "one firm" principle has become increasingly challenging. Yet each one has remained committed, building the integrating channels of communication and forums for decision making to prevent McKinsey from compartmentalizing or fragmenting.

The other key ingredient in making values real is the ability to quantify when and how they have been achieved. In the absence of such measures, the company's strategic goals and operational targets—inevitably qualified and carefully monitored—eventually overwhelm the soft statements of value-laden aspirations. Recognizing this, McKinsey developed client impact measures that put teeth in the firm's commitment to client service. In turn, these efforts led the firm to reevaluate its long-held assumption that its key consulting unit was the office-based, project-driven engagement team. Instead, the firm began organizing around client service teams, a firmwide core of consultants who develop a detailed understanding of a client and its problems over a long period. The objective was to create a group of partners and consultants with a shared collective commitment to serving the clients' long-term needs. In other words, the emphasis on measuring client impact has served to reemphasize the firm values of "one firm" collaboration as a means of providing superior client service.

THE ORGANIZATION AS AN INTEGRATED NETWORK

Traditionally, companies' core organizational decisions have been focused on a few key structural choices. Should we organize around products or markets? Should we structure for efficiency or flexibility? Should we be centralized or decentralized? Behind these managerial dilemmas was an organization that was framed by the classic divisionalized hierarchy that required managers to make such either/or choices. Irrespective of the specific choices a company makes on any of these dimensions, it cannot create the integrated organizational learning capability that we have described in this chapter. The problem lies in the underlying organizational model that forces these choices. In an environment in which jobs are specialized, relationships are formalized, and units are compartmentalized, knowledge cannot flow freely. In the name of efficiency and accountability, the divisionalized hierarchies place a premium on divided accountability at the expense of broader cross-unit collaboration that drives organizational learning.

In contrast, companies like McKinsey, Skandia, Andersen Consulting, 3M, ABB, and others have created an organization based on a framework that can be described as an integrated network (Figure 4.1). This is an organizational model that allows companies to develop distributed capabilities and expertise, link those capabilities through rich horizontal flows of information, knowledge, and other resources, and develop the trust that is required as a glue to hold together their distributed, integrated organizations. It is an organization that is built on two principles that are vital to the development of an embedded learning capability—a structural configuration based on distributed, specialized activities, and a set of relationships based more on interdependence than dependence or independence.

Figure 4.1
The Integrated Network

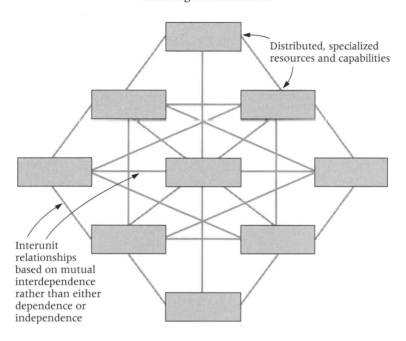

Distributed, specialized
resources and capabilities

Interunit
relationships
based on mutual
interdependence
rather than either
dependence or
independence

The Specialized Network Configuration

To understand the nature of an organizational configuration based on distributed, specialized activities and capabilities, it is useful to contrast it with conventional structures. In the classic centralized hub, for example, resources and authority are centralized to capture the benefits of size and scale, while in the traditional federation structure, relatively autonomous local units retain control over key resources and responsibilities. The problem with both these more common configurations is that they block the kinds of development we have described: The former is not conducive to the blossoming of frontline entrepreneurship, while the latter is unsupportive of cross-unit learning.

In the integrated network organization, the benefits of efficiency are obtained through specialization rather than centralization. Through such specialization, these companies are not only able to capture required economies of scale or critical mass, they are also able to develop focused expertise within each of the specialized units. At the same time, by locating these specialized units in the appropriate markets, these companies are also able quickly and flexibly to respond to diverse external stimuli—the market trends, technological changes, and competitive initiatives that arise in those markets. Thus, McKinsey leads its financial institutions practice from its New York office, not only because it is where the largest client base is but also because this is where its consultants can become immersed in the most advanced practice and develop knowledge and expertise that can become an asset to other firm offices (as was the case for the engagement in Sydney). But, when a New York partner needs to help a client in the oil industry, she will call the McKinsey office in Dallas where the firm's energy practice is centered.

Specialized, distributed activities need not imply permanent ownership of a particular resource or responsibility, and these companies adjust the designation of their "centers of excellence" or "lead units" as the external environment changes or as other pockets of internal capabilities develop. For example, Skandia's expansion into the worldwide annuity-based insurance business has built on what Jan Carendi called a federative organization—a model that closely resembles what we have described as an integrated network. In a context of delegated responsibility and guided by a philosophy that emphasized individual initiative, many of the local companies in Skandia's AFS division quickly began to develop their own particular capabilities and expertise. Wherever Carendi saw such competencies becoming important sources of competitive advantage from which other parts of AFS might learn, he designated the unit as a strategic competence center and charged it with leading the diffusion of its expertise companywide. Because it was the first to develop a unique set of products to be sold through banks, AFS's Spanish company became the competence center for bank product design; the U.S. unit's leadership in developing IT-based systems led it to be designated a competence center for information technology; and the Colombian operation's work in administrative support and back-office functions made it a recognized leader for those strategic competencies. But no lead role was permanent, and each of these units knew that as soon as another subsidiary passed it on the learning curve, it would be forced to cede its prestigious designation.

Integrated Interdependence

The second characteristic of the integrated network is a direct by-product of its configuration based on distributed but specialized activities. In such an organization, interunit relationships can no longer be framed in terms of independence from one another nor dependence of one on the other. Instead, the linkages must be defined in terms of interdependence, not only with the headquarters but also with each other.

Traditional organizations, in which operating units define their key relationships in terms of either dependence or independence, soon find their organizational models impeding their competitive capabilities. Free-standing, largely self-sufficient units are overcome by more coordinated competitors who apply more integrated strategies to pick off such insular operations by cross-subsidizing the losses from battles in one market with profits generated in others. On the other hand, strong dependence of units on the center prevents them from adapting to their different market conditions and responding to their local opportunities. Today's complex and dynamic competitive environment demands collaborative problem solving, cooperative resource sharing, and collective implementation. That is why companies need to build their interunit relationships based on interdependence.

As any family could attest, however, changing relationships of dependence or independence that have developed through a long history of daily interactions is a difficult and sometimes painful process. The classic response of most companies to the growing need for interdependence has been to foster greater cooperation by adding coordinative structures, systems, or incentives. But the results of such a mechanistic approach have generally been disappointing. Independent units have feigned compliance while fiercely protecting their autonomy; and dependent operations have typically embraced the new, more collaborative spirit only to find that it usually implies little more than the right to agree with those on whom they have traditionally depended. Some companies, however, have been able to develop a legitimate and enduring sense of interdependence, not so much by installing structural mechanisms as by focusing on the activities that defined cross-unit relationships. In essence, they have made integration and collaboration self-enforcing by requiring each group to cooperate in order to achieve its own interests.

For example, McKinsey worked hard to overlay a "one-firm" philosophy over its portfolio of seventy local offices. Its four thousand consultants dispersed around the globe quickly learned that not all the knowledge and expertise they needed to serve their clients resided in their offices. Supported by the deeply embedded norms of mutual assistance that the young consultants in Sydney relied upon, no associate hesitated from leaving voice mails or E-mails for other consultants they had never met in distant firm offices where they had few connections. More important, in the McKinsey "one firm" culture, such requests for help would bring responses—from brief messages pointing the questioner to important information sources to commitments to visit the team or the client to help out.

Cross-office personnel transfers were commonplace at McKinsey, either on short-term assignments or longer-term relocations. Firmwide, nearly 20 percent of work was performed by consultants on interoffice loans, and partners universally acknowledged the importance of such moves not only to maintain high-quality client service but also to ensure effective staff development. While some of this selfless willingness to transfer the firm's critical scarce resources so freely could be attributed to the high-minded ideals of client service and human development, another equally powerful motivation was based on enlightened self-interest. From their earliest days in the firm, McKinsey consultants learn that their personal effectiveness and long-term survival depends on their ability to build effective personal networks. Partnership is offered only to those who develop an expertise *and* a network of colleagues who recognize and draw on that knowledge to help their clients. As partners, these individuals transfer the lessons to the people in the offices they lead, recognizing that having maximum client impact usually requires reaching out for expertise, and gaining such cooperation and support demands the development of one's own expertise and knowledge base.

Specialists in Collaboration

It takes enormous effort to create such an integrated network of interdependent specialized operations linked by flows of information and resources, as the experiences of companies described in this chapter confirm. Yet while we have described a series of practical changes we observed these companies undertaking, what may be less clear is that the organization's ability to link knowledge and embed learning often could not begin until there was a more profound shift in top management's thinking. Indeed, in many cases such changes occurred only after there was a basic reconceptualization of the company's business.

McKinsey came to such a major crossroads in the mid-1970s when its unfettered growth of the previous half century began to slow, and its reputation as the premier management consulting firm began to be challenged. Despite the fact that knowledge is the stock-in-trade in this industry, McKinsey had increasingly become compartmentalized into local offices focused on developing deep and enduring relationships with clients rather than exploiting the firm's substantial knowledge assets. Only when a few senior partners recognized that McKinsey had to compete on the basis of "thought leadership" as well as "client relationships" did the firm launch its two-decade-long efforts to build a truly integrated and interdependent organization able to develop and diffuse knowledge rapidly.

And Skandia achieved a similar management breakthrough in the late 1980s when Jan Carendi recognized that his plan to develop a worldwide business required him to reconceptualize the way the company did business. Rather than treat every new market entry as a fresh challenge, he decided to focus his organization on capturing its learning, embodying it in a prototypic model, and transferring it from country to country. Vital to this effort was his ability to create an organization in which information, knowledge, and expertise flowed freely—a learning organization. "We must think of ourselves less as insurance specialists and more as specialists in collaboration," he said.

Today, an increasing number of companies are trying to develop as "specialists in collaboration." It is an organizational characteristic central to a company's ability to develop and diffuse knowledge internally and to make organizational learning a source of competitive advantage.

Ensuring Continuous Renewal:
From Refinement to Regeneration

5

Being singled out as "excellent" seemed to curse many of the companies featured in Peters and Waterman's celebrated book, *In Search of Excellence*. In the decade following its publication, the fate that befell lodestars such as National Semiconductor, Wang, Tupperware, IBM, Avon, Digital, and so many other companies identified as superior performers in 1981 served as warning of the fragility not only of fame but also of success.

What these companies illustrate is a widely recognized phenomenon we call "the failure of success." In many ways, the failure of success exposes us to the soft underbelly of the process of organizational learning described in chapter 4. While it is vital that companies learn from their past experience and capture best practices in policies and routines that allow them to be spread rapidly, such a process can also lead to organizational inertia and inflexibility. Yesterday's winning formula quickly becomes today's conventional wisdom, and without vigilance, can eventually ossify into tomorrow's sacred cow—an unquestionable and unchangeable dogma that locks everyone into doing things "the company way."

The other force contributing to the failure of success is hubris. While a variety of past factors, internal and external, have typically contributed to a company's current profits, growth, and competitiveness, managers of the well-performing company often ascribe the success to their own decisions and actions. To support and protect their successful management processes, they create structures, systems, and staff groups to help monitor activities and keep them aligned with the decisions that ensured the company's current success. This growing internal preoccupation is often accompanied by an external complacency—and, at times, arrogance—as those inside the company begin to believe their own press and start underestimating competitors and treating customers as their captives.

The outcome of this syndrome is well known, having been played out by scores, if not hundreds, of once great companies. The outstanding profitability and growth rates gradually fade, and the company slips into a state of managing what one executive referred to as "satisfactory underperformance." Individually, managers recognize the decline is occurring, and in private discussions they may even bemoan the company's unwillingness to act. Collectively, however, they are unable to challenge the embedded verities or question the systems that protect the status quo. In short, the organization loses its ability to confront reality.

The result of this organizational malaise is that levels of ambition are gradually reduced to match the declining results and rationalization becomes the acceptable substitute for corrective action. This period of satisfactory underperformance can continue for some time, sustained by the accumulated assets of acquired knowledge, customer loyalty, and brand image. Eventually, however, the accumulating weight of inflexible policies and management arrogance drag the company into crisis. Only then is management forced to sit up and ask, "What went wrong?"

Is this grim scenario inevitable? Must a company be faced by crisis to make fundamental change? Is the failure of success some kind of natural law, as inevitable and unalterable as the law of gravity? For those who infer laws from the measure of averages, the phenomenon might well be defined as such since it appears to describe the development cycle of most highly successful companies. But if the definition of a law states that it is inviolable, we can refute the proposition that the law of the failure of success is a law at all, for we have sufficient evidence of companies that have continued to renew themselves quite impressively over time. One such company is Kao, the $7 billion Japanese corporation that has grown from its humble roots as a soap maker to become its country's largest consumer packaged-goods company. It has done so not by focusing on its past achievements but by continuously and constantly challenging each individual—and the organization as a whole—to become the best it can be. In so doing, Kao was able to continue reinventing itself and expanding in order to achieve the dominant position in household cleaners and bleach, capture half of Japan's laundry detergent market, and claim the largest share of the Japanese cosmetics market.

ENSURING CONTINUOUS RENEWAL AT KAO

Founded in 1890, the Kao Company was built on the basis of a simple strategy: produce soaps of equal quality to imported brands but at more affordable prices. The management and employees of Kao Soap Company believed strongly in the rather grand motto they developed for their tiny company: "Cleanliness is the foundation of a prosperous society." For over half a century the company grew, improving its products and building its distribution. Then, in the immediate postwar era, following an explicit policy of imitating and adapting foreign technology and marketing approaches, Kao launched the first Japanese laundry detergent. In following years the company expanded into dishwashing detergents and household cleaners, establishing itself as one of three major Japanese companies that dominated the domestic household cleaning market.

It was not until the 1970s and 1980s, however, that Kao began to pull ahead of its competitors and, more impressively, inflict some humiliating market defeats on the newly arrived foreign interlopers, Unilever and Procter & Gamble. It was in this era, under the two-decade leadership of Dr. Yoshiro Maruta, that Kao developed a management philosophy and organizational capability that wove continuous renewal into the fabric of the company's ongoing activities.

When he assumed the presidency of Kao in 1971, Maruta brought with him a management approach that reflected his deep involvement in Buddhist philosophy. (Indeed, he always introduced himself first as a Buddhist scholar and then as president of Kao Corporation.) At the heart of his beliefs was a principle of human equality that was expressed as a profound respect for the individual. Reminiscent of 3M's core belief system, this philosophy manifests itself in a commitment not only to give employees their own voice but to help them achieve their full potential.

Starting from this philosophy, Maruta introduced a radical concept. He insisted that his managers view Kao not as a soap and detergent company but as an educational institution. He convinced them that the most basic responsibility of every member of the organization was to teach and to learn. As he created his learning organization, Maruta developed sophisticated information systems that allowed managers to capture and process vast amounts of data, adding value and transforming it into usable knowledge as they did so. And he replaced Kao's traditional focus of imitation and adaptation with a new emphasis on creativity and innovation.

Maruta also insisted that the knowledge building and the learning focus on the future rather than reflect on the past. So great was his concern that the organization might become complacent that he discouraged his managers from talking about past achievements or even making historical comparisons. "Past wisdom must not be a constraint," said Maruta. "We must continuously challenge the past so that we can renew ourselves each day." It was this management philosophy that launched Kao into a technology development and business expansion program that grew the company's product line well beyond its traditional soap and detergent roots. In the first half of the 1980s, the company expanded into disposable diapers, cosmetics, and even floppy disks, a product that leveraged the company's expertise in surface science, polymer chemicals, and microfine powder technologies. By the end of the decade Kao was the number one or number two competitor in each of these new market segments.

Unlike most companies, Kao has been able to avoid the learning trap—the widespread mind-set, reinforced by the TQM movement, that convinces managers they are maintaining their company's competitiveness if they continually improve their products and processes. Because they are driving their organizations down a learning curve, they believe the organization is renewing itself for the future. In fact, what it is doing is simply reinforcing and refining its past achievements.

What Kao's experience shows is that true renewal is built not only by ensuring continuous refinement—moving down the learning curve—but also by creating a sense of regeneration—the ability to jump to new learning curves. We saw the same capability in Intel and GE, which have also developed the schizophrenic ability to respect and refine the past while simultaneously attempting to improve upon it. As we examined these and other companies, we identified several common elements that seemed to be vital to this self-renewing capability:

- An internally generated sense of energy that becomes an antidote to the disease of satisfactory underperformance. We call this antidote stretch—an embedded norm and an imposed standard that stimulates each individual to achieve his or her potential and challenges each organizational unit to be the best it can be.

- Organizational flexibility capable of managing the enormous amount of tension generated by the need to balance the conflicting demands of refinement and regeneration.

- An ability to supplement the traditional role of ensuring strategic fit and organizational alignment with another role that ensures strategic challenge and organizational disequilibrium.

CREATING A SENSE OF STRETCH

The organizational complacency and individual timidity that lead to "satisfactory underperformance" are usually supported by structures and systems designed to protect the company's historical source of competitive advantage. The more specifically corporate leaders spell out their strategic priorities and the more tightly they define their business boundaries, the more those deep in the organization feel constrained. These frustrations are compounded by budget-driven goal-setting negotiations that, as described in chapter 3, drive toward the lowest common denominator, usually calibrated in financial terms and focused on a twelve-month horizon.

When Jack Welch took over as CEO of General Electric, he recognized just such an environment in GE. His immediate inclination was to shake employees out of their lethargy, setting the tone in a speech he gave in 1985:

Most bureaucracies, and ours is no exception, still think in incremental terms rather than in terms of fundamental change. For me, the idea is to shun the incremental and go

for the leap. . . . I hope you won't think I'm being melodra-
matic if I say that the institution ought to stretch itself, to the
point where it almost comes unglued.

The size and speed of the changes Welch introduced over the
following decade confirmed his belief not only that the organization needed to be stretched but that he was willing to test the
limits of the elastic. Yet, powerful as his various top-down initiatives were, Welch later conceded that they were unsustainable
unless the culture also began to change from within.

Stretch is not just about top management redefining its strategic vision in more grandiose terms or substituting seemingly
impossible targets for the usual budget objectives. The bigger
challenge is to encourage those at middle levels and in the frontline operations to see themselves and the organization not
through the lens of past achievements or current constraints but
in terms of future possibilities. In other words, it is more about
adapting attitudes and beliefs than about imposing programmatic
change.

The task of liberating the organization from the restrictive patterns of the past and lifting employees' expectations of themselves and of others is not an easy one. In the companies that
were most effective at building stretch into their culture, there
seemed to be a few basic requirements: to create a sense of shared
ambition, to back it with a bond of collective identity (to support
the shared ambition), and to translate both into a feeling of personal commitment and action.

Building a Shared Ambition

Few people want to work for an organization that aspires to be
average. It is part of human nature to want to excel, to be part of
a winning team. Although all good managers recognize this simple truth, surprisingly few have been able to respond to the need
it creates within their own organizations. The problem has usu-

ally been that they have tried to engage their employees intellec-tually through the logic of rational strategic analysis rather than emotionally through the seduction of bold ambition.

At the broadest level, this can be an overarching ambition that gives more personal meaning to the company's long-term objec-tives than the need to achieve a 25 percent return on equity. For Kao, such a shared ambition was instilled into the organization by Dr. Maruta from the moment he became president in 1971 until his retirement in 1994. Maruta wanted every employee to understand that Kao was obligated to develop its technologies and apply them in innovative ways. He created within the orga-nization a winning ambition to create a broad portfolio of prod-ucts that were useful to society and offered good value to con-sumers.

Kao's statement of purposeful ambition stood in sharp contrast to the banality of so many other companies' superficial mission statements printed in annual reports and then ignored by every-one, including those who articulated them. Within Kao, the cre-ation of useful new products became the driving engine of activ-ity and the touchstone for evaluating projects and making decisions. At his retirement, Maruta believed that the company's remarkable record of diversification and growth was due not to some clear-sighted product-market analysis or insightful compet-itive strategy, but to this untiring sense of ambition to harness its technology to improve people's lives. "As a company, we do not spend our time chasing after our rivals," he said. "Rather, by mas-tering our knowledge, wisdom and ingenuity to understand how to supply the consumer with surprise products, we free ourselves of the need to care about the moves of our competitors." It cer-tainly was a more inspiring ambition than selling more soap.

Andrew Grove, CEO of Intel Corporation, also understood the power of shared ambition. After all, Intel had been founded with the bold ambition of replacing magnetic core computer memories with an entirely new product, a memory on a semiconductor chip. Because the product life cycle of a generation of memory products was usually less than three years, Intel soon found itself engaged in a pitched battle with Motorola, Texas Instruments, and dozens of other domestic and foreign companies to become the world's fastest innovator and most efficient producer of memory products.

As Grove eventually discovered, when a company's articulated ambition is too tightly drawn, it can act as a constraint rather than as an energizing force. Through the early and mid-1980s, Intel struggled to maintain its market share and profitability as Japanese memory products flooded the U.S. market at prices that were below U.S. producers' costs. Eventually, Grove acknowledged that Intel's ambition to become a dominant memories company was constraining the organization's ability to aspire to something larger—something that would free it from the destructive competitive cycle in which it was stuck.

In 1986 he made the decision to refer to Intel not as a memories company but as a microcomputer company. He began talking to the organization about becoming the premier supplier of building blocks to the computer industry, a role that he felt could put Intel "at the center of computing." It was a bold ambition for a struggling memories producer that had lost over $250 million in the previous two years, yet it was also a liberating one. It allowed the company to legitimize its gradual exit from memories products and helped people become reengaged around the challenges of making microprocessors that would become the standard for the industry. Less than a decade later, Intel's Pentium chip had become that standard.

Developing a Collective Identity

Setting one's sights on highly ambitious objectives can be an intimidating prospect for many companies and people, since the bolder and more aggressive the goals, the greater the likelihood is that they will not be reached. Yet, acting in concert with others, most individuals develop a sense of courage and commitment they are unable to muster on their own. This is the potential that we saw various companies capture as they generated a sense of collective identity around the ambitious targets of becoming the best they could be.

Kao's open and mutually supportive learning environment, framed by Maruta's concept of the company as an educational institution, created just such a collective commitment to the company's technology and new product development ambitions. For example, the R&D division defined its core activity as "learning through cooperation," and daily work activities routinely transcended the structural boundaries that separated businesses, lab sites, and work units. Current research prospects were discussed weekly in "open space" meetings to which people from all parts of the organization were invited to contribute. Over time the entire R&D group developed a strong yet unspoken sense of shared identity with and responsibility for all of its major projects, regardless of who the formal sponsor was.

However, building a collective identity does not mean losing a sense of individual responsibility, and companies must ensure that collectivity does not degenerate into undisciplined "group think," time-wasting bureaucracy, individual free riding, or other such value-destroying behaviors. Intel is one company that understands this well. It provides an excellent example of how the resulting support of collective group identity must not be allowed to obscure the hard edge of shared individual fate.

In Intel, much of the collective identity comes from the need for mutual support in a business that has always thought of itself as "living on the brink of disaster." That's how company chairman and cofounder Gordon Moore describes it. From its earliest days, when Motorola's 16-bit microprocessor threatened to capture the market Intel created, the company learned it had to act together in a massive focusing of resources in order to survive. What Operation Crush achieved in stopping Motorola in 1978, the Pentium Recall War Room replicated sixteen years later, as the organization eventually mobilized its collective energies to prevent a marketplace meltdown following the discovery of a minor flaw in its flagship Pentium chip.

Grove likes to compare the collective identity that supports such superhuman activity to the spontaneous collaboration emergency workers often exhibit in responding to natural disasters. The company has learned that such collective identity and shared commitment in a corporate setting must also be driven by a sense of urgency, and Intel's leadership has always provided the galvanizing impetus for pulling together in the face of a common threat. Gordon Moore focused the organization's attention on the need for faster technology development and more aggressive production cost management by predicting that the performance of chip technology per unit of cost would double every eighteen months or so—a prediction that became so reliable it was soon dubbed Moore's law. To add more pressure, Andy Grove made the constant fear of competitors and unexpected market developments a hallmark of his management philosophy, which he captured in the motto "Only the paranoid survive." It was from this cultural soup that Intel developed its "emergency worker" capability to join forces against a common challenge.

Creating Personal Commitment

For years Jack Welch drove managers at GE through the strategic planning and operating budget systems he had inherited, setting higher hurdles for strategic qualification and more demanding objectives for annual growth and profit performance. Yet more than a decade after taking over the reins of the company, he came to the simple but profound insight that "the numbers and objectives don't get you there, the people do." Asked what he would do differently if he had to do his time as CEO over again, he responded that he would move much faster, evolving from a traditional control mode to developing a sense of stretch within people. "It's simple," he said. "Budgets enervate. Stretch energizes."

What took Welch a dozen years to learn has become more evident for a newer generation of corporate leaders beginning their transformation processes today. In creating stretch environments, the inspiring visions and imposed objectives can only take you so far. Even the steps involved in building a sense of shared ambition and collective identity are only of value if they achieve their end objective of changing the mentality and shaping the behavior of the individuals who make up the organization.

Intel has taken this value of personal commitment to the company's success and one's own self-development as far as any organization we studied. The company's burning ambition (based on the idea of a business "living on the brink of disaster") and its strong collective identity (driven by the belief that "only the paranoid survive") created an electric environment in which bright, opinionated, and sometimes brash individuals drive themselves extraordinarily hard to ensure the company's—and their own—continued success.

As he tirelessly propagated his fears of unexpected competitive moves or unforeseen market developments ("The price of leadership is eternal paranoia"), Grove was meticulously careful to ensure that the threats were seen as external. Within the company, he wanted to create relationships that were the opposite of paranoid—transparent and open to questioning. Indeed, any employee could demand an AR—action required—of any manager on any issue. The objective was not only to ensure that he and others at the top would not become overly insulated, but also that individual Intel employees would feel free to challenge the conventional wisdom and take the risks that inevitably came with stretch.

But Grove went a step further in urging individuals to become the best they could be. He told his people that they were not Intel employees, but were in business for themselves. He told them they were in competition with millions of similar sole proprietorships around the world. To remain competitive, individuals needed to accept responsibility for their personal self-development and ownership of their careers. He suggested that each individual add value every day, not out of fear of being fired by the company but out of the necessity to meet the challenge of more efficient workers in competitive organizations who could put employees out of work.

Intel's ability to renew itself and to make the right choices in an industry where changes in technological development and industry standards were extremely unpredictable was highly dependent on the willingness of individuals to question existing company strategy and practice and push the organization beyond its comfortable linear development path. It happened in the company's earliest days when despite extreme resource constraints, passionate champions persuaded management to stretch the company support for both MOS and bipolar semiconductor technology. And it continued into the 1990s when spirited frontline debate emerged over the ability of emerging RISC-based processors to challenge the CISC design that Intel had supported, eventually leading the company to authorize parallel-track development of both architectures.

BUILDING ORGANIZATIONAL FLEXIBILITY

The great power of organizations has always been that they provided structure and order to a world that, despite its relative stability, was complex and disorganized. The permanence of most organizational roles allowed specialized expertise and experience to accumulate while the durability of relationships ensured continuity of practice and process. When unexpected change occasionally challenged the function of this rigid model, top management would swoop in to resolve any tensions or dilemmas that were created.

Then things changed. Stability gave way to routine discontinuity and order was replaced by permanent chaos. As the 1980s progressed, the top executives of major corporations found themselves out of touch with the fast-changing environments that shaped their businesses. And even when they were aware of the external discontinuities, they found that their sheer volume overwhelmed their ability to deal with these opposing forces. But the biggest problem was that the rate of change required a much faster response than those at the top could provide. As GE's Jack Welch succinctly put it, "Ultimately, our organization's speed and flexibility is the only weapon we've got."

Nowhere is the need for embedded organizational flexibility more clearly felt than in a company's quest for continuous self-renewal. This vital capability rests on an organization's ability to manage the often conflicting demands of continuous refinement and frame-breaking regeneration—in effect, a kind of organizational schizophrenia. This is an embedded capability that is all but impossible to impose from above. It is evident in Intel's ability to develop the disciplined commitment to drive down a 70 percent experience curve in semiconductor production—squeezing out 30 percent of cost with each doubling of volume—while simultaneously creating the freedom to challenge the very source of that success by creating new generations of products that will make obsolete those with which it is accumulating so much experience and efficiency. Even within the product development process, the conflict between technology-driven solutions and market-led ideas creates a tension that Intel has learned to embrace and manage rather than deny or minimize.

One of the greatest challenges in building an Individualized Corporation is ensuring that the symptoms of organizational schizophrenia do not become pathological. Building and maintaining the structures, processes, and cultural norms to sustain the delicate balance takes a great deal of skill. Among the most important contributory steps we observed were the creation of an organization structure that reflected the multidimensionality of the task, the exploitation of dynamic management practices and organizational processes to give flexibility to static structures, and the development of individual perspectives and skills in a way that "builds matrix in managers' minds."

Building Structural Multidimensionality

Choices concerning organizational design have traditionally been framed in terms of exclusivity: Should we build structures around products, markets, or functions? Should we organize globally, regionally, or nationally? Should we centralize or decentralize control? As they resolved these questions, organizations were created in which there was much greater clarity because the structural choices eliminated, or at least minimized, the complexity.

In the closing decades of the twentieth century, a different set of organizational assumptions has begun to emerge. Companies are quickly recognizing that if they are to retain their flexibility, they cannot afford to eliminate or subjugate whole sections of management perspective or organizational capabilities. Imposing simplicity on a world that is inherently complex comes at a cost, as ABB, for one, recognizes. Percy Barnevik had the vision to decide that ABB should become an organization that was big *and* little, local *and* global, decentralized yet also managed with a degree of central control.

Like ABB, most companies today cannot afford to ignore the fact that they are now operating in extremely complex environments where survival depends on the ability to understand and respond to multiple demands and opportunities. In such a world, a company's organizational structure must reflect, not deny, the complexity of its external environment. To respond flexibly, companies need to develop the range of internal perspectives (in order to understand the environment) and the diversity of resources and capabilities (in order to respond to it).

Kao successfully built such a multidimensional organization. Maruta's strong belief in the individual led him to reject the authoritarianism of hierarchy, preferring instead to represent his organization as a series of concentric circles. Like King Arthur's Round Table, the circular model reflected the egalitarianism and open democracy on which the Kao management philosophy was based rather than a bureaucratic "pecking order."

This organization and the operating reality it helped to create afforded another important benefit. It allowed the company to break away from the functionally dominated model institutionalized by the old organizational chart. Gathered around the board of directors at the center of the circle, business groupings, corporate projects, and geographic units could emerge, taking their place alongside the R&D laboratories and sales divisions that had previously monopolized the top levels of Kao's hierarchy. To emphasize the multidimensionality of Kao's system, Maruta also gave managers multiple assignments. For example, he gave each of the company's R&D heads responsibility for the technical developments in one of the company's business units. In this way he was able to reduce parochial attitudes and break down the biases of focusing primarily on Kao's historically successful soap and detergent businesses.

Intel also evolved far beyond its original functional structure and came to view its organization in much more fluid and multidimensional terms. Matrixed relationships linked product and functional units; shared responsibilities became the norm; and throughout the organization, project teams cut across traditional boundaries, tapping into resources and expertise wherever they were available. In fact, the formal structures and responsibilities changed so fast that many within Intel no longer had a clear sense of permanent organization at all. As one senior executive explained, "We see all organizational forms as transitory. Their purpose is simply to respond to the needs of the time."

One way Intel had embraced flexibility was unusual, if not unique. Its "two in a box" principle made it commonplace for the company to head up a business or unit with two executives with complementary skills. The concept was born of the recognition that Gordon Moore, the technology-oriented futurist, and Andy Grove, the detail-oriented pragmatist, had become a more powerful team than if either one of them had headed the company alone. Eventually, the practice spread to all management levels as a means of stabilizing a transition, jump-starting a start-up, or managing a reorganization. More than anything, however, it became the company's way of broadening the bandwidth of knowledge and experience being brought to a key job, at the same time leveraging the skills and expertise of its talented but often specialized employees. In doing so, it brought a multidimensional perspective to every key decision.

Creating Dynamic Processes

As structures become more multidimensional and transitory, they dilute the clarity of the roles and the stability of authority that made formal hierarchies the major means of defining power and allocating resources in most corporations. As the dominant organizational role of formal structure recedes, managers are relying more on their abilities to create and manage processes as a way to lend the flexibility to their day-to-day operations. (In fact, as chapter 8 will explore, some leading-edge companies are changing the way they think of their organizations—from a hierarchy of tasks to a portfolio of processes.) The particular processes that are central to the task of organizational self-renewal, however, are primarily those that help redeploy scarce resources and capabilities from one set of activities to another—most often from the company's historic line of business to emerging opportunities in different fields.

Having downplayed the importance of structure in Kao, Dr. Maruta preferred to concentrate on developing an organization "designed to run as a flowing system." Building on this image, he employed a variety of other analogies and metaphors to focus managers' attention away from structural images and toward more process-based conceptions of the organization. In what he termed "biological self-control," he described an adaptive model in which ideas, abilities, and resources flowed freely through the organization to where they were most needed. "Just as the body reacts to pain or to injury by sending relief or support to the affected area, so too must the organization respond," said Maruta.

"If anything should go wrong in one department, others should sense the problem and help without having to be asked."

Through such evocative metaphors—and even more so, through the clearly communicated philosophy on which they were based—Maruta created an environment that was highly receptive to more flexible cross-unit initiatives. After years of personal involvement and active encouragement, he found that the organization spontaneously began to find ways to redeploy resources. When an internal plant-efficiency initiative caused a group of workers to become redundant, the surplus workers assembled into a special task force from which flying squads were dispatched to help the company's foreign plants adapt more quickly to the machinery and production techniques being transferred from Japan.

Intel's model of emergency relief teams captured many of the same elements of speed, flexibility, and responsiveness that could also be seen at Kao. Like Maruta, Andy Grove had spent many years creating the environment that fostered such self-organizing initiative. It was interesting to note that as he watched it emerge, he used a similar metaphor to describe it:

> When problems arose, the right experts spontaneously went into action like a bunch of antibodies fighting a foreign substance in the bloodstream. The objective was achieved flawlessly, without a single "manager" ever formally being put in charge.

Building "Mind Matrixes"

In the end, renewal is either propelled or constrained by the capacity of a company's employees to recognize and embrace change, and thrives only in organizations where the narrow perspectives and parochial behaviors of those who live in highly bureaucratic organizations are broken. Thus the core challenge of creating organizational flexibility is to develop individual flexibility, for, as one manager expressed it, it is not so much building carefully designed matrix structures as it is "building a matrix in managers' minds."

Self-renewing companies like those we studied achieved the "mental matrixes" through a variety of human resource policies and management practices that were far removed from the traditional models of personnel development. Rather than encouraging individuals to focus on specialized activities or directing them to pursue predictable, linear career paths, these companies seemed much more intent on exposing key people to as broad a range of perspectives and experiences as possible, and then moving them opportunistically throughout the organization. Their working assumption, contrary to traditional belief, was that specialization did not equal compartmentalization, and that those with a particular expertise should be exposed to various facets of the organization in order to test, adapt, and apply their skills to the company's activities.

Kao, for example, had institutionalized an extraordinarily open management process that exposed individuals to a broad range of ideas and perspectives in the normal flow of everyday decision making. For starters, all employees were encouraged to find out as much as they wanted to know about the company and particularly about how their job fit into the big picture. As a result, all company information was accessible to employees at every level, and anyone could retrieve data on the sales record of any product, the performance of any unit, the status of any new product development, or even the most recent research findings at any of the company's laboratories. "In Kao, the 'classified' stamp does not exist," said Maruta.

Employees were regarded as the company's key strategic resource at Kao, and management's task was "to link them with each other in a union of individual wisdom from which emerges the group's strategy." This deep-seated belief that capable individuals with free access to information provided the best source of creative ideas and healthy debate was reflected in the company's design of what it termed "decision spaces." These were large, open areas at the center of an office floor or research lab, with a conference table, overhead projectors, and whiteboards, where people gathered to discuss and decide on critical issues. The agendas of such key meetings were widely publicized, and anyone with something to offer was welcome to join. Even people walking by would often stop and add their input.

While somewhat more structured, Intel's meetings were equally democratic, and young engineers often found themselves alongside senior executives in key decision-making meetings. As Grove continually insisted, those with "knowledge power" had to be able to deal as equals with those with "position power." This was the only way Grove felt that the status quo would be routinely questioned and successfully challenged. Intel also used unconventional job assignments as a way of meeting certain task needs, shaking up the organization, and developing an individual's bandwidth. In a culture where people gained status and respect for what they knew and achieved rather than where they ranked in the corporate hierarchy, it became routine for individuals to be reassigned to lateral positions or even to jobs several levels lower on the organization chart. As Grove explained, "Careers advance at Intel not by moving up or down the organization, but by meeting company needs." It was a strategy that provided both the company and the individuals employed there with a great deal more flexibility than they had experienced anywhere else.

CREATING DYNAMIC DISEQUILIBRIUM

There are many top-level managers who are great sloganeers for the need to change, yet, deep down, are among the organization's most conservative constituency. But make no mistake— employees quickly see past the slogans and speeches and take their cues from management actions. When the rallying cry for more aggressive technology development stands in contrast to the declining R&D budget; when the commitment to strategic change is negated by continuous rejection of new business proposals; and when the call for employee commitment is followed by a major layoff, initial confusion and frustration among employees can quickly grow into widespread skepticism and cynicism about the leadership's commitment.

The conservative bias observed in many top managers is hardly surprising. After all, they are where they are because they designed—or, at the very least, contributed to—the company's past successes. They wrote the rules of the game and succeeded by playing by those rules. Intellectually, they may accept the need for change, but emotionally—and to a large degree, unconsciously—they yearn for the continuity, stability, and protection of the past they have built. The conservative bias is reinforced by a corporate model that provided those at the top with some powerful tools for institutionalizing existing strategies, policies, and practices. Planning processes produced documents designed to define strategies over three-, five-, and even ten-year horizons; organizational structures, once designed, were announced with great fanfare, published in rich detail, and assumed to reign for years; and individual positions were clearly defined, with responsibilities carefully calibrated so they could be allocated job rating points that allowed them to be locked into a compensation framework. In short, the corporate framework that top management controlled was designed to create a static and stable world that was both predictable and easy to control.

It was in this environment that business school academics and management consultants found an attentive audience for the fashionable notion that top management's role should focus on achieving "fit" and "alignment" among the strategic, organizational, and managerial elements of their operations. It was a powerful notion that provided comforting reassurance to executives searching for clarity and control in a world that seemed bent on upsetting the stable corporate climate they were trying to protect. But a few were distinctly uneasy in such comfortable and controllable environments. Corporate leaders like Grove and Maruta (as well as Welch, Barnevik, and many others) understood that while they must respect the past and build on it, a central part of their role was to decouple the organization from the now-outdated substance of its previous existence and provide individual members with the stimulus and support necessary to leap into a more uncertain future. Going beyond the task of ensuring strategic alignment to providing strategic challenge, these managers complemented their responsibility to maintain organizational fit with a willingness to create organizational disequilibrium. And, most of all, within the turmoil, they were willing to make choices and commitments to the new options and opportunities they stirred up.

Strategic Alignment and Challenge

There can be no question that an organization needs clarity about and commitment to its strategy. Employees at every level must share a well-developed understanding not only of the nature of their company's business but also the reason they are expected to be committed to it and the ways in which the company plans to maintain its competitiveness. In most companies, such strategic focus is ensured through a variety of means, from the formalization of the planning systems to the self-reinforcing spiral of success. The far greater problem is preventing strong strategic alignment from atrophying into strategic inertia.

Andy Grove described the problem well. Despite Intel's spectacular success in setting the industry standard for microprocessors, he fretted: "The more successful we are as a microprocessor company, the more difficult it will be to become something else. . . . We're going to have to transform ourselves again, and the time to do it is while our core business is so strong." It was another revelation of Grove's permanent state of mild paranoia, a quality that prevented him from ever becoming content with being a classic "strategic aligner." He realized that his much larger role was to be a "strategic challenger," ensuring that the organization never became too comfortable with the status quo. One way in which Grove built constant challenge into Intel's activities was to support an approach that became known as "buying options." In an industry where changes in market trends and industry standards were continuous, the ability of a company to constantly adapt its products, technologies, and even its basic strategies was vital. Because it was often difficult to assess the viability of most of the emerging forces, let alone predict their likely impact, Intel's top management learned to back the judgment of those with "knowledge power"—the people closest to the customers or most intimately familiar with current technology with "knowledge power." Often this meant supporting ideas that challenged the company's entire existing approach, but if there was strong commitment to the idea from a group of knowledgeable experts, Grove would agree to back it.

"Buying options" was born early in the company's history when, despite extreme resource constraints, passionate champions persuaded management to back the MOS technology that competed with its founding commitment to bipolar memories. Several years later, supporters of two competing microprocessor technologies both received funding. And the tradition continued into the 1990s, when management permitted a spirited debate to take place over the virtues of the traditional Intel CISC chip architecture compared to the rival RISC design. The company provided a massive amount of funding for a group that wanted to develop the competitive RISC architecture, reevaluating its decisions at every fork in the road. With every decision, however, management was careful to recognize the learning that accumulated along "the road not taken," thus ensuring that the company retained the organizational ability to challenge existing approaches and develop new strategic options.

Maruta, too, saw his role at Kao more as a strategic challenger than a strategic aligner. It began with his broad challenge to the organization to stop thinking of itself only as a manufacturer and marketer of household cleaning products and aspire to becoming an educational institution that developed and applied technology to improve lives and contribute to society. But he did not allow this redefined mission statement to float around in the rarefied atmosphere that inevitably suffocates lofty visions; he continued challenging his organization to live up to its new expectations. At the operating level, Maruta became an active cheerleader for an initiative that began exploring how Kao could employ its knowledge base in oil science, surface technology, and liquid crystal emulsification processes that could be applied to the development of a cosmetics line. And he reacted strongly when old biases within the company combined with conventional wisdom about the cosmetics industry influenced some to propose a traditional marketing-driven strategy for the new products. He urged the individuals involved to go back to the original mission and pursue the company's technology-based, value-enhancing principles. As a result, Kao's cosmetics were the first to be developed and marketed on the basis of functionality rather than image. Emphasizing dermatologically correct skin care, ultraviolet sun protection, and other such functional benefits, the products were a huge success. Within six years they were Japan's second-best-selling line of cosmetics.

Organizational Fit and Disequilibrium

Complementing the need to align individual activity with a common understanding of strategic priorities, managers have long subscribed to the notion that all the elements of an organization—the structure, the controls, the incentives—should complement each other in an elegant organizational fit, which supports the chosen strategy. It is a powerful if somewhat elementary notion, and, up to a point, is inarguably true. But just as companies have found the need to counterbalance the stultifying effect of strategic alignment with constant challenge, so too have they recognized that organizational fit must be offset by occasional disequilibrium.

At Intel, frequent reorganization was a way of life, as was the associated movement of people up, down, or across the reconfigured structural landscape. In this environment, where assets, resources, roles, and responsibilities were constantly being realigned, it was difficult for individuals or groups to become too comfortable or complacent. More powerful, even, was a management style known within Intel as "constructive confrontation." Founded on a series of intersecting team assignments and project activities, Intel's constructive confrontation process was based on the strong belief that those with "knowledge power" were responsible for sharing it with those with "position power." In a style modeled after Grove's own direct approach to debating issues, meetings were conducted in a way that encouraged questioning and argument from all participants. It was by surviving the rigorous debate that raged in these forums that radical proposals worked their way to the top of the company.

Maruta encouraged similar internal debate at Kao in a softer, less confrontational manner that reflected his company's philosophy. In Kao, the prevailing practice was referred to as *tataki-dai*, an operating principle that required individuals to present their ideas to their colleagues at 80 percent completion so that they could be critiqued and developed by others before being locked in as decisions. In addition, Maruta routinely used companywide programs in a deliberate ploy to upset the organizational balance. In the midst of the creative expansion that dictated a simultaneous rollout of new product in cosmetics, disposable diapers, and floppy disks, Maruta inaugurated a major initiative he designated "Total Cost Reduction" (TRC Program), forcing managers to switch from their expansionary focus to a more conservative mode. Four years later, after some major productivity improvement, he co-opted the TCR acronym to become "Total Creativity Revolution." In the metamorphosis, Maruta again put his thumb on the balance, refocusing attention on innovative investments made possible by the cost savings that were generated by the original TCR program. It was a cycle of dynamic disequilibrium to which the Kao organization had become accustomed.

Willingness to Commit

Everything described to this point—creating stretch, building flexibility, ensuring disequilibrium—would suggest that the top manager's role in ensuring continuous renewal is to create an organizational environment characterized by constant turmoil, disturbance, and self-questioning. This is not so. While the objective of having all organization members continually driven to push beyond past achievements and existing boundaries implies a major shift in top management's primary role—from controlling strategic content to framing strategic context—it does not suggest that they abandon their direct strategic decision-making responsibility. Amidst the chaos, someone must be willing and able to make the final choices and commit to the company. When key decisions must be made, the strategic buck still stops on the CEO's desk.

The CEOs of most successful self-renewing organizations have long since given up any idea that they alone must lead the company into new product markets or technology arenas that will drive the next generation of growth. This does not mean, however, that they have delegated all responsibility in this vital area to the ideas and initiatives of those deeper within the organization. One of the most powerful levers corporate leaders control is the ability to make substantial resource commitments ahead of clear opportunities. By creating a form of "supply side economics," they build a tension that is resolved only when new opportunities are developed to meet the committed investment.

This was precisely the approach taken by Kao's Maruta in an effort to strengthen the renewal efforts of his company's household-products business, which had been struggling for years with its internationalization program. After Kao's decade-long joint venture with Colgate was dissolved, Maruta decided that stronger action was called for. In a bold move, he authorized the acquisition of Andrew Jergens Company, a Cincinnati-based soap, shampoo, and body lotion manufacturer, outbidding seventy other would-be buyers. It was a move designed both to focus his managers' attention on the vital U.S. market and to develop their commitment to the internationalization of their excellent products. He took a similar risk when process technologists discovered that Kao's development of new chemical agents, surfaces modification technologies, and microfine powder processes for concentrated detergent products and cosmetics gave it a technology base that could provide it with a competitive advantage in the development of computer floppy disks. To jump-start the business and to provide impetus for this radical proposal for corporate renewal, Maruta authorized the acquisition of Didak, a Canadian-based floppy disk manufacturer. With superior technology, the company launched a successful attack on the U.S. market, achieving $10 million in sales in its first year.

Top management's intervention and commitment was also required in calming the chaos that often resulted from the pot stirring at Intel. While Intel's process of constructive confrontation tended to winnow out all but the best ideas with the strongest internal support, it was still possible for competing initiatives to make it to the top. And although Intel's established practice of "buying options" sometimes led to the support of both initiatives, in the end, top management usually had to decide which initiative to back. After spending more than five years agonizing over the CISC/RISC dilemma, for example, Grove decided to back the traditional CISC technology. It was a decision made in keeping with a philosophy he described as "letting chaos reign, then rein in the chaos." Although such decisions often involved bet-the-company kinds of risks, there was no doubt that in the end, it was Grove's job to make the bet. "When you come to a fork in the road you have to decide which direction to take," he said. "Otherwise you risk hitting the divider."

MANAGING SWEET AND SOUR

If managers seek to avoid the failure-of-success syndrome, they must give up the ideas of simplifying and stabilizing the organizational environment to make it more manageable. Simplicity and stability are the parents of complacency and inertia, which lead companies into the treacherous territory of "satisfactory underperformance."

Continuous self-renewal is built on the tension that develops between two symbiotic forces—the need for ongoing improvement in operational performance as provided by continuous rationalization, and the need for growth and expansion as generated by continuous revitalization. Rather than seeing these forces as mutually conflicting, companies like Intel and Kao trust them to be complementary, leveraging each to drive and energize the other. The rationalization process focuses on resource productivity, and through a process of continuous refinement of ongoing activities, ensures that current assets and resources are used effectively. Most companies understand the need for such processes, and the vast majority of their structures and systems are designated to drive it. Such widespread measures as return on net assets, value added per employee, inventory to sales ratio, and time to market for new products are all designed to calibrate productivity performance gaps and focus organizational energy on closing them.

The first thing that distinguishes self-renewing companies from the pack is that in addition to driving for continuous improvement they also seek radical leaps in cost reduction and productivity improvement. 3M's stretching target to increase the rate of new product introductions by 50 percent is a classic example. So too is Intel's constant drive to double its microprocessors' processing capacity per unit of cost every eighteen months. And GE's goal of increasing companywide operating margins from 10 percent to 15 percent is another example. The paradoxical secret that self-renewing companies have learned is this: Stretch targets are often actually more achievable than traditional incremental objectives, precisely because they force radical rethinking of current operations.

The second distinguishing feature of self-renewing companies is that the rationalization process is seen as a continuous activity, not a one-time cleanup job. As many of those who have undertaken major reengineering programs have found, after the trauma of massive layoffs and plant closings the organization often breathes a collective sigh of relief and, exhausted and somewhat traumatized, goes back to business as usual. In self-renewing companies, by contrast, the need for year-on-year productivity growth and continuous cost reduction never goes away—as is obvious to anyone visiting a Kao manufacturing operation or observing an Intel design team. More than imposed targets, continuous improvement becomes the shared commitment of all the organization's members—an ongoing personal responsibility, not an occasional corporate program.

The complementary part of the renewal process is revitalization—the challenging and changing of the existing rules of the game that usually results in the creation of new competencies and businesses rather than the refinement of old ones. It is the antidote to a rationalization-dominated mentality that has measured the success of a corporate transformation program by its impact on profitability and head-count reductions. Yet such heady short-term gains are often due to the simple truth that the company is most profitable in the latter stages of its growth cycle—from the time it stops investing until the time the business dies. To ensure organizational continuity, the focus on costs and resource productivity must be accompanied by an equal commitment to growth and opportunity creation.

Creating new opportunities is different because it inherently involves much greater risk than refining old opportunities. Embarking on an aggressive growth path is like trying to ride a tiger—if you fall off, you probably won't survive. So while rationalization calls for grit and tenacity, revitalization demands courage and commitment. Intel's seven-figure investments required in every new generation of chips—what Andy Grove calls "investing in science fiction"—call for such boldness. So too did Maruta's willingness to back Kao's equally risky move into cosmetics, a field in which the company had no prior experience but which represented a huge (and ultimately fruitful) new opportunity.

In most companies, the processes of rationalization and revitalization are seen as conflicting and even mutually exclusive. Rationalization is the "sour" medicine few managers enjoy administering or taking, yet many companies have been locked into this pattern for years. Imprisoned by an ever-shrinking cycle of identifying performance gaps, making painful cuts, and watching competitors catch up and surpass them, these companies were caught in a self-fulfilling prophecy, creating the need for yet another round of cutting (and another and another). Still other companies have tried to ignore rationalization by focusing on the "sweet" agenda of revitalization—gathering the management team for an off-site visioning exercise, planning the growth agenda for the new millennium, and so forth.

In self-renewing companies, however, these two processes are seen as symbiotic—the continuous rationalization process provides the resources and legitimacy for revitalization, which, in turn, generates the hope and energy required to sustain the grueling rationalization process yet again. In other words, these companies have learned the recipe for mixing sweet and sour, a formula that takes some of the bitterness out of cost-cutting programs and adds a little zing to the company's growth activities. A classic example of sweet and sour cooking is Intel's survival of its near-death experience during the memory-chip bloodbath of the mid-1980s. In the midst of a vicious price war, its competitors resorted to immediate layoffs and shutdowns. Intel opted for another path. In the first round of the crises, it introduced a "125 percent rule" that required that everyone from factory worker to CEO work ten extra hours a week for the same pay. Redeploying resources redeployed from the beleaguered memories business to the more promising microprocessor activities, the company worked to ride out the storm. As the pricing free fall continued, the company was forced to close plants and lay off some people, but it minimized the effects by adopting its "90 percent plan" by which all employees took a 10 percent pay cut while continuing to work the ten extra hours each week. To raise more capital to invest in microprocessors, management even sold 20 percent of Intel's equity to IBM.

In the late 1980s, Intel emerged from the trauma, having preserved most of its valuable human resources and redeployed them to new, more promising businesses. It maintained the lead it created in microprocessors during this difficult period by continuing to manage "sweet and sour"—continuously balancing rationalization and revitalization. It is a process that requires everyone in the organization to drive for continuous improvement at the same time as they question and challenge the sources of their past success. For Andy Grove, that meant being willing to turn his back on the memories business—the business on which Intel was founded and to which he himself had been so personally committed.

Part 3

BUILDING AND MANAGING THE INDIVIDUALIZED CORPORATION

Shaping People's Behaviors: Changing "The Smell of the Place"

6

The best way to introduce the topic of this chapter is to relate it to an experience that will be familiar to most readers. Recall the time when you last found yourself in the middle of a big city like New York, Mexico City, or Calcutta, in midsummer. The heat was unbearable and the humidity stifling. The snarled traffic belched exhaust that hung in a thick smog around you, while you were being pushed and jostled by a crowd of people who were all as tired and frayed by the oppressive conditions as you were.

Now recall the last time you found yourself on the slopes of Aspen, in a New England village during the fall, or in the forest of Fontainebleau in France in early spring. The day is bright and clear, fresh, clean smells surround you, and there is an invigorating crispness of the air.

With those images in mind, you begin to grasp what is perhaps the core of the problem being faced by many large companies. Most of them have created within their organizations the oppressive, energy-sapping environment of downtown Calcutta in mid-summer. This environment, in turn, saps all the initiative, creativity, and commitment from their people. That is why they cannot renew themselves.

There is one central truth in corporate transformation that is often overlooked: Companies cannot renew their businesses unless they first revitalize their people. But even among those focusing on this objective, the challenge often appears to involve changing people—changing their attitudes so as to change their behaviors. Indeed, one can find few underperforming companies today that are not in the middle of such a program to "change employees' mind-sets."

Most of these programs are misconceived and misdirected. Companies are highly unlikely to create entrepreneurship, for example, through inspiring corporate commitments, powerful incentives, and training workshops. Our belief, stated from the earliest chapters of this book, is that all the entrepreneurship and creativity most companies need are already present within their organizations but are stifled and debilitated by an oppressive internal atmosphere. Just as no amount of inspirational speeches or individual coaching will make you feel highly motivated and energized in Calcutta in midsummer, neither are the programs designed to change mind-sets likely to rekindle the entrepreneurial spark plugs in companies that have created their own internal downtown Calcuttas.

Rather than focusing on changing individual behaviors, the more important challenge is to change that internal environment—what we call the behavioral context—that in turn influences people's behaviors. To reshape these behaviors, managers must transform their behavioral contexts. Instead of downtown Calcutta, they must transform their companies into Fontainebleau forest in spring.

THE SMELL OF THE PLACE

What we call the behavioral context, one manager described as "the smell of the place." "Walk into any office or factory," he said, "and within the first ten minutes you will sense the smell of the place. You will sense it in the energy and the hum of work, you will see it in the eyes of the people, in how they walk and talk. You will sense it in a thousand small details all around you." Whether described as culture, climate, or context, it is the smell of the place that prevents companies from creating the capabilities of entrepreneurship, learning, and self-renewal that we have described in the previous three chapters of this book.

To describe the typical behavioral context in most companies and to illustrate its pernicious effects on people's behaviors, let us return to the story of Westinghouse. The story is particularly appropriate because, over the last fifteen years, this company has declared victory in the battle for change—not once but three times, each time to stumble and begin the process all over again.

The first victory, described in chapter 2, was announced by Robert Kirby, when, after a massive restructuring spanning the entire second half of the 1970s, he announced an end to the company's history of "unpleasant November and December surprises." "We have killed the beast," said Kirby, referring to the complaint of slack internal controls that had earned Westinghouse the label of an "accident-prone company." "We are at the beginning of an era of uninterrupted growth," he declared in 1981.

By 1983, however, a sharp deterioration in results lead to a fresh round of changes in the company's strategy, structure, and systems. By 1987, as return on equity topped 20 percent, exceeding that of arch-rival GE, Danforth had declared victory again, announcing Westinghouse's entry into the "winner's circle"—a small group of elite corporations enjoying a reputation of consistently superior financial performance and managerial excellence.

But the victory was short-lived. After another setback in 1988, new CEO John Marous announced his vision of elevating Westinghouse "from the good corporation it is today to a great corporation," triggering one more round of changes in the company's portfolio and organization. By 1989, with double-digit sales growth and a net profit of nearly a billion dollars, Marous's vision appeared close at hand, and this time it was not management but *Fortune* that declared victory. Comparing Westinghouse's recent performance with that of GE—to the advantage of the former—*Fortune* labeled its cover story on the company "Respect, At Last."

Two years later, however, soon after incoming CEO Paul Lego took over, a massive mess at Westinghouse Credit—perhaps the biggest "surprise" in the company's history—plunged Westinghouse into a 1992 loss of $1.7 billion and marked its stock down to half the level two years earlier. Unable to ride out the storm of criticism, in early 1993 Lego was replaced by Michael Jordon, who announced "a new beginning."

The bottom line of the story is stark. At the starting point of the tale, Westinghouse and GE were comparable companies, seen and spoken of as worthy rivals. At that time, Westinghouse was about 60 percent the size of GE in terms of annual revenues. At our 1995 end point, GE is nine times larger (Figure 6.1). In a two-decade search for corporate renewal, three successive generations of top management presided over a massive decline of a major American institution.

Figure 6.1
Comparison of GE and Westinghouse
(Revenues, 1976–1996)

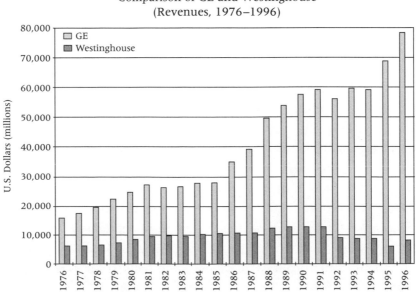

The Corrosive Context

There are many reasons for Westinghouse's inability to achieve durable transformation. However, the roots of its corporate sclerosis lay in its behavioral context. And Westinghouse is not an isolated exception. Like Westinghouse, many large companies have, over the years, developed a behavioral context that, while superficially benign, has had a corrosive effect on the behaviors of their members.

Only by explicitly recognizing the central characteristics of this inherited context and understanding how it affects management perceptions and actions can those who want to revitalize their organization replace its most pernicious qualities with others more conducive to genuine and durable growth and renewal (Figure 6.2).

The first element of the traditional company's managerial context is what we term *compliance*—a characteristic that was important, even vital, in the postwar years when many companies diversified their activities into scores of inviting opportunities. As they began their rapid expansion into a diverse range of new businesses and markets, most found an urgent need to have widely dispersed employees complying with common policies and uniform practices so as to prevent powerful centrifugal forces from pulling their organizations apart. The classic military model of line authority that dominated the formal relationships between managers and employees ensured that those deep in the organization would follow the direction set by the leaders.

But while this widespread contextual norm ensured unity of action at a time when the key management challenge was to choose among competing opportunities, it could also degenerate into a pathological form. Here, inflexible procedures and authoritarian intolerance of dissent prevented outmoded policies from being challenged and shut down meaningful debate on top-down directives.

Ultimately, it was not the policies themselves but the effects they had on the day-to-day behaviors of people that made it so

Figure 6.2
Pathologies of the Inherited Context

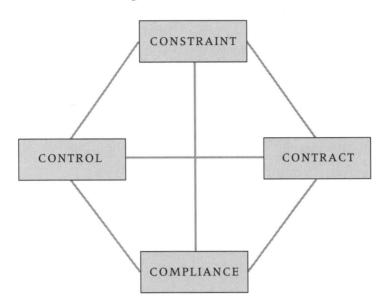

difficult for these companies to sense early-warning signals and to correct problems before they became disasters. One of the most damning public charges laid against Westinghouse CEO Paul Lego was that "there was no one to challenge him." Although for years many within the company could see the impending collapse at Westinghouse Credit Corp ("Even the guys in the mail room were asking when there would be a write-off," said one former executive), Lego apparently remained unaware of the mounting problems. In a culture in which authority and order had quashed dissent, top management had become isolated from day-to-day operations. This tradition, built into the company's behavioral context in the 1970s, was carried into the 1990s when the massive unexpected losses in financial services were followed by major unforecasted problems in environmental services.

The second common characteristic of the managerial context in large modern corporations is *control*—again, an organizational characteristic that had allowed companies to expand operations rapidly and efficiently in an earlier era. This deeply rooted norm that characterizes classic hierarchical relationships had been greatly strengthened with the introduction of the divisional organizational structure. Corporate-level executives were willing to delegate substantial responsibility to a new level of general managers only if they had the mechanisms to hold them accountable. Strongly influenced by powerful corporate staffs, most companies had developed sophisticated capital planning and operational budgeting systems to establish top-down control throughout their organizations.

While such systems proved highly effective in allocating funds and driving ongoing performance, they eventually contributed to a deterioration in interpersonal relationships. The objective-setting and forecasting processes often degenerated into a game-playing exercise between adversaries, and the monitoring activity frequently became an excuse for an increasingly powerful corporate staff to intervene in the operations of frontline managers. Once again, the impact on the frontline managers' behavior was to make them defensive and risk-averse.

In a management group dominated by engineers, tight controls had long been a central characteristic of Westinghouse's management style. The principle of tight control was firmly reasserted after a decade-long experiment with freer management had led to a serious performance decline in the mid-1970s. But the shorter and tighter leash placed on people led to constant complaints about the haggling in the planning and budgeting processes. The highly sophisticated system was based on company-imposed estimates of capital costs and cash flows that became an unending source of debate, and many felt the system was driving them to achieve short-term results at the expense of long-term business development. About the only topic on which there seemed to be general agreement was that the tightly administered processes were consuming an enormous amount of management time and energy.

In the traditional large-organization model, another strong influence on attitudes and behavior was the nature of the relationship between the corporation and its employees—one based on contract. This was a characteristic born of legalistic biases that became greatly strengthened by two more recent organizational trends: the highly incentive-leveraged compensation systems that reinforced the notion of a financially based relationship between the company and its employees, and the massive restructuring, rationalization, and redundancy programs that underlined the fact that this relationship could be terminated at any time.

Although initially the implicit or explicit contract between employee and employer served to define expectations and give the relationship clarity and stability, it eventually led to a formalization and depersonalization of the way in which individuals felt about their companies. As widening compensation differences fostered resentment and increasing terminations bred fear, people began to distance themselves emotionally from an entity they felt had betrayed them. More and more, they felt like employees of an economic entity, and less and less like members of a social institution.

The familylike relationship that once dominated the Westinghouse culture began fraying in the 1970s when CEO Robert Kirby resolved to revive the company's sagging fortune with a massive program of layoffs and divestitures that cut the total workforce by 30 percent in three years. However, more than the layoffs per se, what destroyed the family bonds at Westinghouse was the manner in which the layoffs were carried out. Any illusion of an emotional relationship between the company and its people was soon squashed by Kirby's colorful but disconcerting statement that he would fire his own mother if she wasn't producing the expected results. Twenty years later a never-ending series of layoffs and divestitures had reduced the workforce from 200,000 in 1974 to 54,000. A once strong sense of company pride and loyalty had almost entirely vanished.

The other dominant characteristic common to the behavioral context of many modern corporations is *constraint*. As companies expanded and diversified, top management found it increasingly important to develop clear, focused definitions of corporate strategy as boundaries within which those with delegated responsibility could operate. Particularly in an environment in which opportunities for expansion exceeded most companies' ability to finance them, such constraint was helpful in preventing diversification from becoming unmanageable or from wantonly dissipating precious resources.

Eventually, however, as broad strategic objectives were specified in detailed strategic plans and translated into specific portfolio roles for different businesses, the strategic constraints became confinements and the operating boundaries became barriers. Managers of businesses classified as mature began to think of themselves as mature, becoming risk-averse and innovation-resistant. The strategic process became constraint that not only affected how managers could act but ultimately how they could think. Constantly bombarded by strategic visions, roles, goals, challenges, and priorities, frontline managers retreated into a passive mode far removed from the spirit that had once powered the organization's growth engine.

The degeneration of this once legitimate element of management context was well illustrated in Westinghouse. As a way to get control over operations that "had gone totally hog-wild," CEO Robert Kirby introduced the concept of strategic business units (SBUs) and imposed strict discipline through the company's highly touted planning system, Vabastram. Initially, Vabastram (Value Based Strategic Management) educated an engineering-oriented management team to a more financially oriented view of the world, improving the discipline of Westinghouse's investment process. Ultimately, however, top management's blind faith in Vabastram deprived the business units of all flexibility and creativity. It reshaped behaviors, both within individual businesses and across them as each of the thirty-seven SBUs focused on its own business, attempting to maximize its own return on allocated equity. It also reshaped the frontline managers' relationships with top management. Vabastram provided top managers with the data they needed to decide whether to continue to invest or to sell off the business. As they made seventy divestitures between 1985 and 1987 alone, the message to the organization was clear: deliver current performance or your unit will be sold.

When Michael Jordan was named to replace Paul Lego as CEO of Westinghouse in mid-1993, he identified not only the massive challenges he faced in reviving the company's sagging operating performance and restructuring its damaged strategic portfolio, but also the huge task of transforming an internal management culture that he described as "a throwback to the 1950s." Although a 1950s-based culture was ideal in the postwar era in which a company's opportunities exceeded its ability to fund them, in an environment in which innovation, responsiveness, flexibility, and learning had become vital sources of competitive advantage, a management context framed by compliance, control, contract, and constraint had become a liability, not an asset.

THE CONTEXT FOR RENEWAL

The portrait we have drawn of the behavioral context in large corporations is not a flattering one, and to some extent is a caricature. Although there may be only a few companies in which all four elements of the context have deteriorated to the degree we have described, there are equally few that have emerged untarnished by any trace of such pathologies.

Why has the historically evolved behavioral context proven to be so debilitating? The reason is that as all four core characteristics atrophied, they drove management to become passive, compliant, and inward-focused—captives of their glorious past rather than explorers of a brave new future. The only enduring antidote to this pervasive disease of corporate sclerosis is to build a different behavioral context, one that drives a company toward continuous self-renewal rather than refinement of existing capabilities and defense of current positions.

As we explained in chapter 5, self-renewal capability requires companies to manage the essential symbiosis between the present and the future, to balance a cycle of "sweet and sour." In companies that have developed an ability to continuously renew themselves, management is acutely (and often painfully) aware that there can be no long-term success without short-term performance, and that short-term results mean little unless they contribute to achieving the long-term ambition. They consciously manage rationalization to provide the resources and credibility to take the risks and make the investments that revitalization requires, and they use revitalization to create the energy and sustain the hope needed to engage in rationalization processes that are often demotivating and exhausting.

To better understand how these two forces—the yin and yang of continuous self-renewal—are most effectively managed, it is instructive to compare the approach of Westinghouse with that of Andersen Consulting. In the former, the importance placed on resource productivity was reflected in the waves of layoffs and divestitures that invariably followed the appointment of each new CEO. During the 1980s, successive campaigns under three CEOs resulted in the sell-off of well over a hundred businesses and the reduction in the total number of employees by more than 60 percent. During the same period, however, the company also acquired more than fifty businesses, representing the main thrust of the corporate leaders' commitment to revitalizing a company whose internal growth and profit they saw as being constrained by its portfolio of mature operations.

At the same time, Andersen Consulting's top management was also struggling with the need to renew its operations in the cauldron of one of the world's most fiercely competitive and fast-changing industries. But in both approach and outcome, the contrast with Westinghouse was dramatic. Through a constant struggle to keep itself ahead of the "commoditization envelope," the firm eventually emerged as the largest consulting company in the world.

While Westinghouse management was acutely aware of the need to rationalize and revitalize, their efforts to deal with the company's low productivity and slow growth were dominated by a top-down, project-driven approach that was reinforced by each incoming CEO's determination to make an enduring impact. As a consequence, however, sequential initiatives and actions ran into the back of each other, and revitalization programs that had invested in new resources and activities were often swept aside in subsequent rationalization drives. (A classic example was the acquisition of Teleprompter and Unimation in the early 1980s, signaling a major commitment to cable television and robotics as sources of future growth; within five years both acquisitions had been sold off as part of subsequent rationalization waves.)

In contrast, the senior partners of Andersen Consulting made renewal the outcome of the day-to-day choices and actions of people at all levels, while top management's role was to develop, nurture, and protect that organizationwide process. In fact, the very emergence of a consulting practice within the guts of a traditional audit and accounting firm was not the product of a grand vision of Andersen's top management but rather an outcome of the initiative and commitment of Joe Glickanf, then an ordinary partner, who, in 1951, built a copy of an advanced design computer that had been developed at the University of Pennsylvania. When he presented it to the senior partners with a vision of what it could mean for the future of business systems, they voted to give full support for further development of this activity. That support, in turn, led to a contract for implementing an automated payroll system at GE in 1952—believed to be the first commercial application of a computer—and, therefrom, to the birth of a new business that has since evolved into the $5 billion consulting arm of the firm.

To develop the kind of management understanding, belief, and commitment that drives this kind of self-renewal process from within, companies have to build a very different kind of behavioral context than the norms of compliance, control, contract, and constraint that hobbled Westinghouse and so many other large organizations. As we examined the management processes in successful, self-renewing companies, we identified four common features of their behavioral context. We came to describe them as discipline, support, trust, and stretch (Figure 6.3).

Each of these elements has played a role in earlier chapters: Chapter 3 explored the need for discipline and support in stimulating individual initiative; chapter 4 focused on trust as a key facilitator in establishing horizontal integration and organizational learning; and chapter 5 detailed stretch as an energizing force that drives continuous self-renewal. Unlike compliance, control, contract, and constraint, these dimensions do not rely on authority relationships or management policies to influence behavior. Rather, each dimension becomes integrated into the flow of a company's ongoing activities and is reflected in every aspect of daily corporate life.

Figure 6.3
Management Context

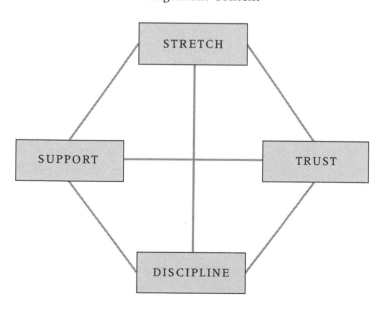

Discipline is more than compliance to directives or conformity to policies; it is an embedded norm that makes people live by their promises and commitments. It is this characteristic of Andersen Consulting to which competitors refer when they deride the firm's consultants as "Andersen androids"—although the term could also reflect a touch of professional envy of the uniformly high standards of quality, accuracy, and thoroughness to which Andersen's associates hold themselves.

The firm has developed its disciplined approach over many decades, protecting it through careful socialization of all new employees. The principal means the company employs is the intense education and development processes previously described. The first session, in a series of educational activities involving a thousand hours of training over a recruit's first five years, is a six-week program that the new employee is more likely to compare to boot camp than to an executive education course. Eighty-hour weeks, demanding assignments, and a strict dress code are all intended to create a norm of discipline as much as to provide training in new analytical techniques.

In self-renewing individual corporations, if discipline substitutes for compliance, then support replaces control. In such companies, the relationship between bosses and subordinates is defined by characteristics of coaching, helping, and guiding. Sure enough, this feature is immediately identifiable at Andersen in the roles the firm's partners play. All partners are expected to commit themselves personally to the development and support of the firm's associates, a responsibility they describe as "trusteeship." At the time of recruitment into the firm, each new associate is assigned a counseling partner who accepts responsibility for meeting with the new member every six months to discuss performance, career interests, and development needs. In addition, project managers also provide associates with feedback and support, evaluating performance every three months and coaching them during the project on an ongoing basis.

But the context of support applies to more than just the vertical relationships previously dominated by control. It also frames the horizontal linkages among peers. In part, such horizontal relationships are far more naturally developed within a partnership-based practice—as we saw in the case of McKinsey—but, as in McKinsey, they are also strengthened in Andersen through a variety of mechanisms, including industry and functional specialist networks as well as strong firmwide norms that require people to voluntarily seek and provide assistance to colleagues.

The third characteristic of a self-renewing behavioral context is trust. People who trust one another rely on each other's judgments and depend on reciprocal commitment. While the relationship between a company and those who work for it is inevitably defined by a mixture of contractlike agreements—both explicit and implicit—and by a more organic familylike emotional bonding, historically most companies have drifted toward the contractual end of the spectrum. In the process, individuals and organizational units are motivated to protect their self-interests and maximize their side of the contract, resulting typically in an increase in adversarial relationships and an erosion of the two-way commitment.

Trust is reflected in and reinforced by transparency and openness in a company's organizational processes and by a sense of fairness built into its management practices. Andersen Consulting has nurtured these characteristics over many decades of consistent action. One manifestation of trust is the way it compensates its partners—a thorny issue in all professional service firms. At Andersen, a partner's fraction of the profit pool is determined by the number of "units" he or she is awarded, based on evaluation by other partners not only on quantifiable dimensions, such as business generated or studies directed, but also on judgmental criteria such as practice leadership, associate development, and teamwork contributions. All partners receive a listing of all others' unit allocations, and although there is an official appeal process, nobody in the firm can recall it ever being used.

Finally, stretch—the fourth dimension of the behavioral context of self-renewing companies—is the liberating and energizing element explored in chapter 5 that raises individual aspiration levels and encourages people to lift their expectations of themselves and others. In contrast to constraint, which limits perspective and restricts activities, stretch induces a striving for more rather than less ambitious objectives.

At Andersen, a firm commitment to a principle first articulated by its founder provides the basis for such stretch. "The client deserves our best," said Andersen, the man, and it is a dictum that has since driven Andersen, the firm, to continuously pursue better ways to meet client needs and stay at the forefront of its rapidly evolving business. It is this firm commitment to client service that justifies all the evening and weekend hours for its individual members, as well as the large investments the firm makes—often ahead of demand—to build the capabilities to meet emerging market opportunities. From its first investment in a new computer to its first steps in internationalization, every major step in the growth of the firm has come from this concept that the client always comes first, regardless of efforts or personal sacrifice.

It is difficult to convey the enormously demanding nature of a context that is shaped by these four elements. The combination of the hard-edged characteristics of stretch—which drives people constantly to strive for more ambitious goals—with discipline—which embeds the norm that all promises must be kept—defines an environment that is as energizing as it is taxing. Even the softer contextual norms of support and trust create an organizational culture in which reciprocal obligations place enormous demands on the individual. And while we have described each of these attributes separately, they are neither independent nor static. Indeed, it is in the interplay among these dimensions and in the co-evolution of each that the whole dynamic of self-renewal is created.

Thus far, we have described "the smell of the place" companies need to develop to help them become Individualized Corporations and have contrasted it to the Calcutta-like context that framed the behaviors of the "organization man." But companies like Andersen Consulting that have developed the characteristics of discipline, support, trust, and stretch over long periods of stable leadership—3M, Intel, and Kao are other examples—are, in many ways, the exception. While few companies suffer from all the contextual ills of a Westinghouse, most share at least some of its pathologies. For managers in these companies, the big question is: How can they begin to reshape the contextual attributes in their organizations? What does it take to replace compliance with discipline, to move from control to support, to create stretch in place of constraint, and to build trust where now there is only contract.

THE TRANSFORMATION OF PHILIPS SEMICONDUCTORS

From our observations in several companies, we firmly believe that it *is* possible for a determined management team to transform the behavioral contexts of their companies—and to do so in a reasonable period of time. Such change was at the heart of the transformation of companies such as AT&T, ABB, and Corning. But where we saw the most radical example of context change was in the semiconductors division of Philips, a prime example in describing the core dimensions of the new behavioral context and in illustrating the kinds of management actions that are necessary to create it.

In 1990, when Philips's performance fell off the precipice, the business that held the doubtful honor of contributing the most to the company's misfortunes was the semiconductor division. While the division had been losing money and market share continuously for three years, it chalked up a staggering loss of $300 million that year, on sales of about $1.8 billion. A detailed group-level investigation at that time also revealed the strategic position of the business as "hopeless." In a scale-intensive global industry, the division was ranked tenth, and its products were positioned in low-growth and highly cost-competitive segments. Furthermore, the business was extremely investment intensive, with average industry R&D spending of 15 percent of sales, and capital investment of 130 percent of depreciation. To meet such investment needs, the group report concluded that a company needed at least a 6 percent global market share; Philips's semiconductor division had less than 1 percent.

Internally, the group-level manager responsible for the semiconductor division described the situation within the division as "catastrophic." Management conflict was rife. The powerful head of the only profitable business unit strongly and openly disagreed with the chief executive's strategy of investing heavily in a new technology; two of the four business unit managers did not even speak to each other; and relations between the line management and the relatively autonomous R&D group were extremely strained. At the operating level, there was what one manager described as "complete paralysis" made worse by the ultrapolitical environment.

In March 1989, in response to the division's deteriorating financial and competitive situation but prior to its sudden downturn, its chief executive was replaced by Heinz Hagmaister, the head of one of the four business units. Unlike his predecessor, who had limited company and industry background, Hagmaister had spent his entire three-decade-long career at the semiconductor division of Philips, rising through its technical and marketing ranks. His agreement with the group management had envisaged nursing the business back to the break-even point by 1992, and during his first fifteen months he had focused his attention on trimming R&D and controlling expenses.

All that changed in May 1990 when the financial results of 1989 were announced. Stung by the reaction of the financial markets to the unexpectedly large losses, the group appointed Jan Timmer corporate president. The new CEO's first act was to demand more urgent action to control costs. Over the next three months, the divisional management team prepared a plan to reduce personnel by 20 percent, to cut the R&D budget by more than 50 percent, and to close several facilities, eliminating the related products from the line.

During the next three years, as these plans were gradually implemented, the division experienced a remarkable metamorphosis. Despite the recession-plagued industry situation in which both demand and prices for its products declined, Philips's semiconductor performance steadily improved: 1989's loss of $300 million shrank to a deficit of $150 million in 1990; the business broke even in 1991; and profits of $200 million, $250 million, and nearly $350 million followed in the next three years. Clearly, such a rapid and dramatic turnaround in a business that had been declared dead was worth examining more closely.

As we tried to understand the sources of this radical improvement, it became clear that a variety of factors had all contributed to the change—from a major financial restructuring to a strategic decision to focus on niche products, from an overhaul of the divisional management team to a significant change in the global marketing organization— Yet, as we talked to a large number of people at all levels within the division, each highlighted a profound change in the internal environment as the primary reason for the division's dramatically reviving fortunes. Indeed, it is from one of them—a young engineer in the company's power semiconductor plant in Scotland—that we first heard the term "the smell of the place":

> What matters most is the smell of the place has changed. I now enjoy coming to work. It's not one thing, but overall, it has become a very different company.

What changed "the smell of the place"? How were discipline, stretch, trust, and support assimilated into the day-to-day operating context of the Philips's semiconductor division? What were the key management levers?

As we explored these issues with people from various levels and divisions of the organization, we identified a set of key events and actions that separately and collectively helped bring about the change. Some of those actions (for example, a benchmarking exercise) influenced more than one dimension of context (both stretch and discipline). In other instances, while some people identified one particular action (such as the introduction of a new accounting system) as contributing to the development of one dimension (such as discipline), others described the same action as contributing to another dimension (such as trust). Overall, what emerged from those discussions was a broad sense of how a focused and consistent set of managerial actions can shape the behavioral context, and how behavior context, in turn, influences action.

Instilling Discipline

The vast majority of the people we spoke to identified an increase in discipline as a key element of the transformation. Beyond the normal expectations of timely reports and financial goals, the employees described a broader change, one that led to the development of a strong norm of "management by commitments." As one manager explained:

> We now send our samples to customers on time, or at least we try our best. We phone back if we have said we would, and we turn up in meetings on time. If something has been decided in an earlier meeting, we don't reopen the issue. In fact, meeting your commitments has become kind of an ego issue; it's not just about the inventory, cost, or revenue targets but, more generally, about doing what you promised.

As we pieced together the different explanations of why and how discipline became a key element in the context of the division, three key factors stood out: the establishment of clear standards and expectations; the implementation of a system of open and fast-cycle feedback; and consistency in the application of sanctions (see Figure 6.4).

Figure 6.4
Emergence of Discipline as a Key Element in the
Context of Philips Semiconductors

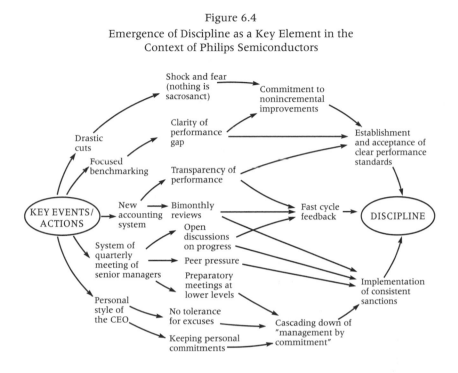

Clear standards. The establishment of clear standards required not only the development of expected performance markers but also an acceptance of and commitment to them. Detailed benchmarking against a key competitor, according to most managers, was the vital tool that helped establish clear standards of performance throughout the organization. While earlier benchmarking exercises—using general data provided by external consultants or from public sources—had allowed the division's managers to develop some standards, it was the quality and detail of the data collected in the 1990 benchmarking process—this time conducted by division managers against data obtained directly from a leading competitor—that made it much more credible. The highly disaggregated and fine-grained information prevented debates on authenticity and comparability, while management's realization that the business might be divested led to an emotional commitment to bridging the visible performance gap. The reality of the performance gap was further reinforced by the data generated by a new cost accounting system that was detailed enough to establish specific responsibilities, expectations, and standards at the level of individuals responsible for small, disaggregated units.

This represented a major change in the company's philosophy of accountability. Historically, in Philips, those who could convince top management that a particular activity was of vital long-term strategic importance could exempt that part of the business from the unpleasant cuts demanded in the name of improving productivity. By contrast, in identifying his targets for the drastic cuts of 1990, Hagmaister had firmly discredited the concept of "strategic importance" by retrenching activities such as R&D spending that were earlier considered sacrosanct. The fear generated by the threat of retrenchment coupled with the embarrassment caused by unflattering benchmark comparisons played a key role in creating an emotional commitment to achieving the new performance standards.

Fast-cycle feedback. Discipline in the organization was also strengthened through activities that increased the frequency and quality of internal feedback. The introduction of a new accounting system provided management with the means to do so. First, by eliminating certain kinds of information needs, the new system allowed rough weekly figures to be compiled by the following Tuesday, generating detailed results on shorter time frames more immediately. The chief executive and the chief financial officer closely monitored these results and personally followed up on unanticipated outcomes.

While the accounting system improved the frequency and timeliness of feedback and the demand for explanations, the introduction of quarterly performance review meetings involving all senior managers from the key units of the business encouraged openness, honesty, and candor in the review process. Hagmaister's personal style (in his own words, "calling shit, shit") helped shape these meetings into an open review of all performance aspects. As the intensive peer-review process caught on, norms were established to discredit overtly political or counterproductive behaviors. By creating similar meetings within their own units to prepare for and discuss the outcomes of the senior managers' meeting, most managers helped institutionalize the new norms of candid and honest feedback throughout the organization.

Consistent sanctions. The third contributor to the development of management discipline was the consistent application of sanctions. Hagmaister established a norm of dealing quickly and firmly with underperforming units and their managers. Replacement of a number of key managers including the chief financial officer and the powerful heads of two of the largest business units lent credibility to the slogan "no excuses" and helped managers face up to the difficult decisions of reducing head count. At the same time, careful review by the corporate human resource group prevented arbitrariness in the process, and well-publicized reversals of two openly political dismissals helped establish the norm of fairness and consistency.

Creating Stretch

> Another major change is how we think about targets. In the past, everything was 5 percent. If anyone proposed changes bigger than that, he was immature, he didn't know his business. Now, if you propose a 20 percent cut in inventory, you are a bit embarrassed because someone else is shooting for 25 percent. That too has become a part of life—how far can we go? And that makes the challenge game!

This comment by the manager of the semiconductor division's largest manufacturing facility in Holland highlighted another key change in its behavioral context: the creation of stretch. Again, while a variety of explanations were offered about how such an environment came to be so strong within Philips Semiconductors, three key attributes were identified by almost everyone: the estab-

lishment of a shared ambition; the emergence of a collective identity; and the development of personal significance in the turnaround task (Figure 6.5).

Figure 6.5
Emergence of Stretch as a Key Element in the
Context of Philips Semiconductors

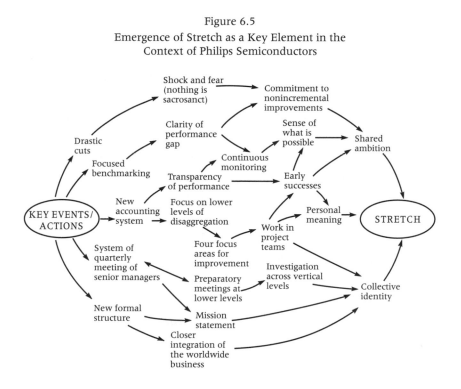

Shared ambition. The same process that led to the establishment of clear standards also helped build a sense of shared ambition within the organization. Beyond the emotional commitment
to achieving highly stretched goals aimed at bridging the recognized performance gap, by 1992 managers had begun to believe
in a future of profitable growth and a reputation for excellence in
specialized market niches. The visible celebration of some early
successes played an important role in converting the desire for
survival into a shared ambition for the future.

Another important contributor was the collective decision to
create an early quarterly senior managers' meeting to focus the

entire organization on four key performance objectives: to reduce the development time for new products (the "time-to-market" project), to cut the cycle time from order entry to product delivery (the "make-to-market" project), to shorten customer response time (the "customer satisfaction" project), and to prune the unwieldy list of fifteen thousand product offerings (the "portfolio choice" project). While each of the four broadly constituted project teams established performance standards and monitored improvements, a large number of unit-level teams implemented the specific tasks. Within six months, some fairly dramatic performance improvements led to a growing confidence that Philips could match, and in some areas, surpass, "best-in-class" performance levels. The transfer of best practices and shared work in the project teams resulted in the spread of success throughout the organization, feeding a new optimism that was validated by improving financial performance, culminating by late 1992 in a fairly widespread shared ambition for excellence.

Collective identity. Soon after Jan Timmer became Philips's president, Hagmaister convinced him of the need to separate the semiconductor business from the company's electronics components division in which it resided and make it a free-standing, independent division. The organizational separation helped managers of the semiconductor business develop a greater sense of collective identity. Said Hagmaister: "We needed to find our own way to do things together."

The split also led to a stronger and tighter integration across the new division's functional operations. In the past, the sales organizations responsible for semiconductor products also had to handle the company's other electronics components. Furthermore, this sales force reported to Philips's local country managers, further diluting the ability of the semiconductor management to control them. After the split, the semiconductor division formed its own specialized national sales organizations, which were consolidated under a marketing manager in the divisional headquarters. Similar specialization was also achieved in the product development units. Freed from the distraction of other activities and diverse reporting relationships, these dedicated units could build stronger links among themselves while the resulting horizontal integration also helped build a collective identity.

Finally, although Hagmaister was skeptical about the need for a mission statement, he agreed to develop one only at the insistence of his management team. Jointly created and adopted by the fifty senior managers in early 1991, the statement describing the semiconductor division's strategic priorities and organizational values seemed to have little effect on the organization until the company's performance improved about a year later. As performance approached the vision, it became a source of pride, and debates on issues from resource allocation to new product development priorities were increasingly resolved by reference to the mission statement. By the end of 1992, it had become a catalyst of common action and collective identity within the organization.

Personal meaning. As the four priority projects cascaded down from the senior managers' quarterly meetings, the broad targets established by the fifty top managers were allocated and translated into specific action. The business unit meetings broke down the targets for each product group, and the product group meetings translated the targets for each factory, development team, and marketing group. As a result, more and more individuals not only had the benefit of focused targets but also had a clear picture of how his or her own tasks contributed to the overall performance of the company. Most of the people interviewed believed that the explicit "line of sight" between their own work and the company's priorities created a sense of personal "ownership" that gave meaning to each individual's work. This association, in turn, created the motivation for stretch at the individual level.

Establishing Trust

Perhaps the single most visible difference between Philips Semiconductor in 1989 and Philips Semiconductor in 1993 was that the people within the organization had begun to trust one another. The difference was indeed dramatic. In 1989, managers in one business unit had been discovered advising a customer not to deal with another business unit that was soliciting an order for a very different set of products. In contrast, by 1993 the various business units were collaborating actively—for example, creating a shared CAD/CAM system and jointly developing new products. Growing trust was seen as a key contributor to this spirit of cooperation.

In describing how such trust was developed, we identified three key contributors: the higher level of perceived fairness and equity in the company's decision processes; the broader level of involvement in core activities; and an increase in the overall level of personal competence at all levels of the organization (Figure 6.6).

Figure 6.6
Emergence of Trust as a Key Element in the
Context of Philips Semiconductors

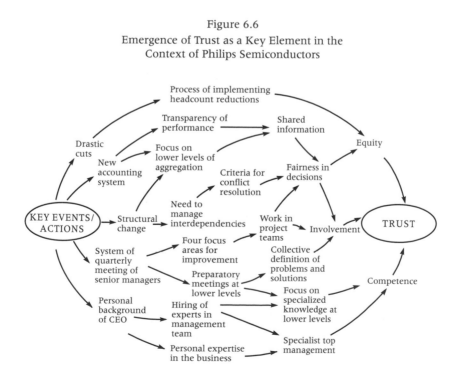

Fairness and equity. The first and most important factor in creating a sense of fairness within the organization was the process for managing the 20 percent reduction of personnel. Because of the unprecedented extent of the cuts, most expected highly politicized decision making based on the relative political clout of each unit manager. Instead, decisions were made in collective meetings, based on objective data of benchmarked performance and business priorities, and no subsequent changes were made in "corridor deals."

The change in the internal structure of the division also contributed to the development of a sense of fairness. In early 1990, Hagmaister restructured the business units within the division around end-user industries rather than around technologies. This change increased the level of interdependence among the units, particularly in the areas of technologies and manufacturing. The need for greater coordination led to the creation of new forums that provided a sense of internal equity—both real and perceived to the dispute resolution process.

Involvement. Teamwork within and across units had increased considerably within the division. The formation of numerous project-groups between 1990 and 1992 dramatically increased the number of people working on core management issues. Furthermore, the system of cascading quarterly meetings not only served to involve more people in decisions that affected their work; it also broke vertical and horizontal barriers that had previously constrained both participation and information access. For example, regular meetings of the fifty top managers spanned five management grades and all functions and business units—a significant departure from the earlier practice of specialist meetings involving only those at the same level. Even when decisions ran counter to the interests of individuals, they were exposed to the broader rationale and had the opportunity to advocate their positions. By enhancing the transparency of outcomes, involvement increased the perception of fairness in the process, thereby raising the level of trust.

Competence. Himself an engineer and industry expert, Hagmaister believed that a high-technology company like Philips had to be managed by specialists, not generalists (like his predecessor). As a result, the two new business unit managers he appointed had specialized expertise in the industry's technologies and production respectively. In turn, these managers brought in more people with experience in the specialized production process and increased the level of specialization in the company's dedicated sales organization. Overall, by increasing the level of competence, this upgrade in specialist skills also contributed to increasing trust. As one manager described it:

> Trusting someone in the bar is different from trusting someone in the business. Ours is a high-tech outfit. I need someone at the other end who knows what I am talking about. I can spare him a CAD/CAM guy, if he desperately needs one, but I must be sure that he can spare a logistics expert if I need one someday. . . . It is easier to build cooperation among people who know the technical aspects of the business.

Developing Support

In identifying support as an element of the new context within Philips semiconductors, the managers we spoke to most often pointed to their increased access to company resources located outside of their own unit, and the less control-focused and more help-oriented senior management role that was allowing additional freedom of initiative at lower levels. Collectively, this greater availability of resources together with increased autonomy and more help created an environment that supported rather than constrained lower-level initiatives and entrepreneurship (Figure 6.7).

Access to resources. Because of conflicts within the top management team and a structure that emphasized the independence

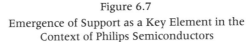

Figure 6.7

Emergence of Support as a Key Element in the
Context of Philips Semiconductors

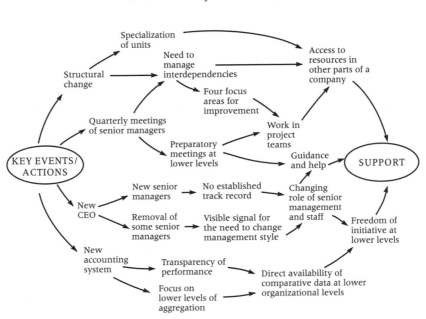

of each business unit, cross-unit cooperation had historically been limited within the semiconductor organization. As a result, for example, each unit had adopted a different IT system, with different CAD/CAM software that prevented access to each other's design libraries.

However, the division's new market-oriented structure required more interdependence, and the changes in personnel increased the number of technically competent individuals with both the interest and the ability to access resources in different parts of the company. A shared CAD system, for example, led to the development of a collective CAD library that all could use. The consolidated sales organization allowed salesmen in one country to use the literature or an order control system developed by another. And the new integrating forums helped a product development team in the United Kingdom to use the services of an expert in Germany. It was through such access to resources that managers on the front lines could operate effectively in the new, more decentralized mode.

Autonomy. Hagmaister's strong commitment to radical decentralization was the primary driver of increased autonomy in all the frontline units of the organization. Again, key appointments triggered changes throughout the organization. When he found that two of the old-time authoritarian business unit managers were blocking the drive for decentralization, Hagmaister replaced them with two technical experts. With limited general management experience, but an inherent preference for decentralization, these two managers played a major role in creating more freedom throughout the units they headed.

The new accounting system also contributed to the decentralization by providing faster, more reliable operating information by product group, in contrast to the earlier system, which only provided profitability estimates at the level of the business units. The new system allowed frontline managers to identify problems more quickly, and with a clearer picture of what was going on, senior managers felt less need to interfere with day-to-day operations. As one manager said, "The rigor of the new system allowed better control. That, in turn, reduced the need for backseat driving."

Guidance and help. The greater horizontal cooperation required by the proliferation of meetings and project teams opened up access to resources and advice from other units and established a norm of mutual help across the division. This cultural change was also reflected in the evolving role of senior management. The replacement of older generalists with younger specialists in the management team communicated a clear message: We are moving from a control focus to a coaching model. One of the new business unit managers explained:

> I see my role as that of a coach—helper, supporter, teacher. I have to influence the overall strategy, and we have made some progress in specializing units to better use our resources. I have to play a role in coordinating across those

> units. But beyond that, my job is to help and guide, to provide
> advice, and to protect my people.

This trend was reinforced by the substantial change in the role
of the company's historically powerful central staff groups. With
their control over the division's information flows almost
destroyed by the large open-meeting format, and their power
substantially curtailed by budget cuts that shrank their size by
over 40 percent, the staff groups could no longer control line
managers as they once had. Newly appointed heads of the
finance, logistics, and human resource functions also supported a
philosophy that said the staff must work for the line rather than
the other way round. Collectively, these developments led to a
new relationship in which the legitimacy and influence of a staff
group depended on the extent of support it was recognized as
providing to line managers.

FROM CONTEXT TO BEHAVIORS

It is worth emphasizing again that many different factors contributed to the transformation of Philips Semiconductors. Without
a $500 million restructuring charge, which helped the division
write down its assets and lighten its balance sheet, the transformation would not have been possible. Without a sharp cutback on
its product portfolio, together with a strategic reorientation from
large-volume commodity-type chips to specialized niche products,
the improvement in financial results would not have taken place.
Similarly, without a 25 percent head-count reduction, the division's cost structure would not have been brought in line with
competitive levels.

Yet, what we have described at Philips goes far beyond the kind
of financial, strategic, and organizational restructuring that
becomes the preoccupation of most managers in their search for
revitalization. The reason so few companies achieve the dramatic
yet durable performance improvement that Philips Semiconductors

managed is clear: In the final analysis, businesses cannot be renewed unless people are revitalized, nor can process reengineering work unless behaviors are changed.

The power of the very different behavioral context we have described in this chapter comes from the internal tensions that exist among the four foundational characteristics on which it is built. People learn to operate in an environment that is, on the one hand, highly disciplined and demanding yet, on the other hand, also trusting and secure; an environment where expectations are stretched, ambitious, yet within a setting that is supportive and nurturing. And it is in the resolution of these complementary yet often contradictory forces that the organization develops the energy and direction to drive its self-renewal process.

In the end, therefore, the power of the behavioral context lies in its impact on the behavior of individual organization members (Figure 6.8):

Figure 6.8
Management Context and Individual Behavior

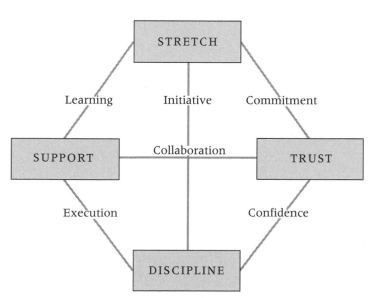

- The ability and willingness of people to take *initiative* is rooted in the tension between stretch and discipline: the former serving as the source of energy and the latter converting that energy into tangible and time-bound action. Stretch without discipline leads to daydreaming, while discipline without stretch locks the company into an ever narrowing spiral of refining existing operations without the courage to make a creative leap.

- Similarly, it is the combination of trust and support that motivates cooperation and *collaboration*. Trust makes cooperation desirable; support enables individuals to convert that desire into action. Each is a necessary element of the organizational glue, but only in combination do they create the sufficient condition for integrating the disparate actions of dispersed people.

- Beyond initiative and cooperation, renewal also requires some other kinds of behaviors in people—an openness to *learning*, the courage of *self-confidence*, the willingness to *commit*, and the ability to *execute*. It is the same four attributes of context that, in different combinations, provide the enabling conditions for each of these behaviors.

These are precisely the kinds of behaviors that the new context triggered in Philips. And while the various financial and strategic restructurings largely yielded one-time relief, it was the commitment to these new behaviors that helped the division turn around a "hopeless" business into the single most profitable part of Philips's global operations.

Our study revealed numerous instances of the ways in which newly stimulated behaviors resulted in important consequences. In a small plant in Scotland, for example, throughput time had fallen from sixty-two days to twenty-four days through the cumulative effect of worker-initiated changes to redesign equipment layout, to improve the handling of intermediate products, and to restructure and reallocate tasks. The changes were proposed and implemented by shop-floor employees despite the fact that they led to staff cuts and reduction of overtime payments.

In another instance, the division created a highly successful new business with a chip for car phones that came about from an intensely collaborative effort of its historically antagonistic German and British operations. The German unit had specialized expertise in designing telephone components; the British operation had a long history of developing products for the automotive industry. In earlier times, the two units would have mounted parallel and competitive efforts to develop products for an emerging market, such as car telephone components, over which both could claim jurisdiction. In 1991, however, the two unit managers came to an agreement whereby the British unit would take the lead in the development of the new product, drawing on its understanding of the automobile market, while the German unit would contribute by allocating several of its engineers with telephone expertise to work full-time on the project for six months.

Quality of Management

Herein lies the essence of the battle for corporate renewal: It is, ultimately, a battle over individual behaviors. As detailed as our descriptions of the contextual dimensions may seem, and as intricate as the charts may have appeared, these representations of the changes in the semiconductor division may have actually obscured the complexity and the challenge inherent in winning the battle of behaviors. What the charts and explanations fail to convey is the importance not only of what was done but how it was done. For example, it was not so much the benchmarking process, the quarterly meetings, or the new accounting system that contributed to developing discipline in the semiconductor division; it was the way senior managers used those tools and activities that was key. The benchmarking process was merely a tool that confronted managers with the size of their performance gap; it took concerted management action to convert it from a mere set of numbers to the division's widely shared norm of striving to meet tough standards. Similarly, Hagmaister replaced imposed compliance with embedded self-discipline by the way in which he used the new accounting data to elevate routine performance reviews into management education sessions that preached the importance of fulfilling commitments. Ultimately, it was not the tools and initiatives but the quality of management in applying them that contributed to the establishment of discipline as an established behavioral norm.

The same is true of each of the other elements of context described. Trust, for example, is an organizational characteristic that is built only slowly, carefully, and with a great deal of time and effort. And here again, management cannot be separated from the message. At the end of the day, through his day-to-day actions and by being who he was, Heinz Hagmaister inspired people to trust him. His willingness "to call shit, shit" on the one hand, and to constantly demonstrate his personal fairness and openness on the other hand, was inseparable from all the systems and processes he put in place and the concrete actions he took to re-create trust as an overall aspect of the company's behavioral context.

It has, by now, become a cliché to claim that people are the key source of a company's competitive advantage. But, in a literal sense, this cannot be true in the long term—at least not since slavery was abolished. The only way people can directly serve as the source of a company's competitive advantage is if they are exploited—in prison factories or sweat shops. In companies that pay a fair wage commensurate with the individual competencies of their employees, the real source of competitive advantage lies in their context—in the internal environment that allows people to individually and collectively create far more value than they could if they were employed elsewhere. If, as is often asserted, the key function of management is to help ordinary people produce extraordinary results and if, as demonstrated in this chapter, the behaviors of people in a company can be so radically changed by changing the internal behavioral context, then shaping that context is undeniably the principal task of managers and the best measure of a firm's quality of management.

Building Organizational Capabilities: The Company as a Portfolio of Processes

7

The fundamental objective of management in the Individualized Corporation is to shape the behaviors of each employee to do what 3M has long described as "stimulating ordinary people to produce extraordinary results." Creating the right behavioral context is the most basic requirement in building and managing a company that stimulates people to take initiative, to collaborate, and to develop the confidence and commitment to continually renew themselves and the organization. But can a company actually organize itself to embed "the smell of the place"? How should it be structured? What roles must managers at different levels play? How would work get done in such a company?

In talking to groups of managers about the Individualized Corporation, one of the most predictable questions is, "What kind of structure do you need to create it?" This is a natural reaction from a generation of managers who found that structural adjustment led to new strategic capability—the creation of product divisions made possible diversification, the installation of international groups drove overseas expansion, and the invention of strategic business units (SBUs) led to tighter product-market focus. Interestingly, though there were commonalties among our sample of companies in terms of frontline entrepreneurship, cross-unit learning, and continuous self-renewal, and even in the dimensions of behavioral context required to develop these capabilities, there was not a common structural form that seemed to trigger or even follow the transformation to an Individualized Corporation. Indeed, many companies seemed to have created entrepreneurial self-renewing learning organizations in spite of their structures rather than because of them.

MAKING BUMBLEBEES FLY

When scientists studied the body weight and wing structure of the bumblebee, they concluded that according to all known principles of aerodynamics, it should not be able to fly. But somehow it does, and once again the laws of nature declare victory over the laws of science.

In the course of our studies we witnessed a number of corporate bumblebees—companies whose organizations were so complex and ungainly that by any stretch of modern organizational theory, they should have collapsed under the weight of their own bureaucracy. And yet they were flying—and, in some cases, soaring.

Take the case of Canon. From its humble roots in the midrange camera business, over the last two decades Canon has achieved a remarkable rebirth by creating a highly sophisticated dynamic growth cycle. This delicately balanced iterative process is based on Canon's ability to first exploit its expanding portfolio of businesses to build new technological and functional capabilities, then use its ever-growing portfolio of capabilities to enter new businesses. This strategy has allowed Canon to progressively expand its operations beyond cameras to calculators, photocopiers, laser printers, fax machines, and computers, each time taking on and humbling larger, resource-rich competitive incumbents.

Not surprisingly, Canon incorporates many of the organizational characteristics of the Individualized Corporation. Like ABB, ISS, and 3M, Canon is built on the foundation of frontline entrepreneurship. In fact, the company's remarkable history of product proliferation started in the late 1970s only after the company's newly appointed president, Ryuzaburo Kaku, reversed his predecessor's highly centralized practices and initiated a process of radical decentralization. That broke the company into small units. He then assigned clear profitability targets to each unit together with ambitious growth objectives, and gave them both the resources and the freedom to devise their own ways to achieve those goals.

To ensure that the knowledge and expertise built within the different business units were integrated and leveraged across the company as a whole, Kaku created three powerful organization-wide systems—the Canon Development System (CDS), the Canon Production System (CPS), and the Canon Marketing System (CMS). Supported by the Global Information System for Harmonious Growth Administration (GINGA)—a high-speed digital communications network developed at a cost of twenty billion yen (approximately $90 million) to interconnect all parts of the company—the main function of CDS, CPS, and CMS is to supplement Canon's efficient vertical communications structure with a lateral one that fosters direct information exchange and knowledge transfer among managers across businesses, countries, and functions on all operational matters.

Canon also represented one of the finest examples of continuous renewal. Its record of creating major new businesses was second to none among the companies we studied; in fact, it is perhaps second to no other company in the world. Yet while growing new businesses, it also continuously drives down costs and refines its products in order to strengthen its position in its existing activities (a classic illustration of the "sweet and sour" process described in chapter 5).

Canon's formal structure, however, appears to be a modern-day variation of the old functional organization that has all but disappeared in the seventy-five years since Alfred Sloan introduced the divisionalized structure to replace it. Canon's Management Committee includes, in addition to the company president, the chairmen of CDS, CPS, and CMS, essentially the leaders of the development, manufacturing, and marketing functions. Each headed a committee that, in turn, supervised a host of small units responsible for development, production, and marketing for the camera, business machines, and optical products businesses. In other words, unlike GE, whose operations are aligned primarily along the business axis, or ISS, where the operations are grouped primarily along a geographic axis, Canon's focus resulted in an integration driven primarily along functional lines. While most managers around the world have deemed a functional focus unworkable in large, diversified companies, Canon has somehow managed to make the bumblebee fly.

In structural terms, 3M, too, is a bumblebee, but of a very different kind. Around the world managers have been delayering and destaffing their organizations to create the benefits of entrepreneurship, integration, and renewal. 3M's capabilities in each of these dimensions have evolved from its outstanding ability to leverage individual initiative and organizational learning to create a portfolio of sixty thousand products and a hundred technologies.

Yet a superficial examination of 3M's corporate structure would suggest that such frontline entrepreneurship should be impossible. Beneath the CEO is the sector organization, headed by sector presidents. Under the sectors are the groups, led by group heads. Then there are the divisions, supported by divisional management teams, and beneath them the departments. Finally, reporting to the departments are business units and project teams, led by people like Andy Wong. And this entire superstructure operated under the watchful eye of a large and powerful corporate staff. This is precisely the kind of organization that GE's Jack Welch had described as being "a ticket to the boneyard in the 1990s." Yet, somehow, 3M appears to have found a way to make its bumblebee fly.

ABB, too, is a structural anomaly—an organization built on a complex global matrix that, according to common wisdom, cannot work for a large, worldwide company. Over the last decade, companies like Digital, Owens-Corning, Citibank, Xerox, and IBM have all tried and then abandoned this difficult structural form after just a few years of frustrating experience. The matrix, they forcefully argued, tends to make a company slow, inflexible, and bureaucratic under any circumstances. But when managers are separated by the barriers of distance, language, time, and culture, it becomes completely unworkable.

Yet, ABB is a classic global matrix, framed by strong business and geographic managements. And despite such a structure, ABB has managed to grow from $17 billion in revenues in 1988 to $34 billion in 1995, reduce its head count in Europe and North America by fifty-four thousand people while building up a forty-six-thousand-person organization practically from scratch in the Asia-Pacific region, and assume industry leadership by stealing market share from such formidable competitors as GE, Hitachi, and Siemens.

The only thing common to the organizational structure of these three companies is that they are each completely unfashionable. None of these bumblebees, in theory, should be able to fly, yet while others fret about wing structure and body weight, these three are gathering more honey than the rest of the hive put together. What became clear to us as we examined companies like Canon, 3M, and ABB is that our obsession with structure is misplaced. What the managers of these individualized corporations were focusing on—consciously or unconsciously—is a set of core processes that is supported by a different set of management roles and relationships.

This is not to suggest that structure is irrelevant, for it clearly is not. But it is better thought of as a framework within which companies can develop the organizational processes and management roles and relationships that are at the core of their competitive capability. We think of structure as the organization's anatomy, an important but insufficient model for understanding how the living organism works. Equally vital is a thorough understanding of the organizational physiology, the processes and relationships that ensure the financial and informational resources that are the company's lifeblood. And finally, we must understand the organization's psychology, the culture and values that shape the attitudes and beliefs of organization members.

To further understand how these elements fit together to create a high-performance Individualized Corporation, we will dissect the operations of a single company, ABB. But first, an important proviso. We describe ABB's organizational development merely as an example, not a model for all companies. As Canon and 3M illustrate, different organizational forms work for different companies. Each entity must define its own version based on its strategic imperatives and organizational history. There are, however, some core processes that are likely to be common to all. These are the dimensions of the emerging organizational model on which we will focus.

THE ABB ORGANIZATION

A Snapshot of ABB's Organization

Structurally, ABB is built around a global matrix (Figure 7.1). At one level, the company can be viewed as a $34 billion global behemoth. But CEO Percy Barnevik prefers to describe ABB as a federation of 1,200 small national companies spread across the globe.

Each of these frontline companies is quite small—on average employing about two hundred people and generating about $50 million in revenues. Most of them, in turn, are divided into four or five profit centers, each employing about fifty people and generating between $10 and $20 million of revenue. But under Barnevik's principle of radical decentralization, each company is structured and treated as a distinct business and, whenever possible, as a free-standing legal entity. The basic objective of creating such a large number of small entities is to ensure that employees will lose "the false sense of security of belonging to a big organization" and will develop "the motivation and pride to contribute directly to their unit's success." In other words, the whole architecture has been designed to work with the grain of individual identity, pride, commitment, and anxiety, thereby shaping people's behavior at work.

Figure 7.1
ABB's Matrix Organization

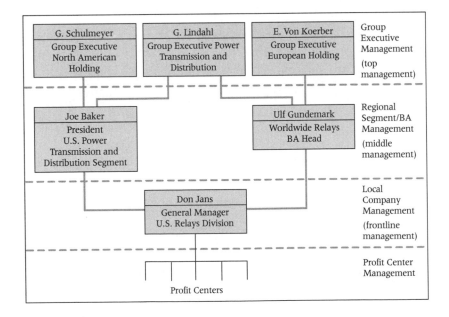

Between the frontline of these 1,100 little companies and the top management in Zurich is the level of regional segment/business area managers, who form the geographic and business arms respectively of ABB's global matrix. Ulf Gundemark, introduced briefly in chapter 2, was the head of the relays business, one of the business areas (or BAs in ABB terminology) within the power transmission and distribution sector. Joe Baker headed that sector for North America. As head of the relays business in the United States, Don Jans reported to Gundemark on the business axis and to Baker on the geographic axis. In turn, Baker and Gundemark reported to Goran Lindahl, head of ABB's power transmission and distribution sector on the business axis, and to their respective regional managers on the geographic axis.

In contrast to the eight or nine layers of management in its predecessor companies, in ABB there is only one intermediate level between the Group Executive and the 1,100 frontline company managers. And, in keeping with the principle of radical decentralization, staff support at this level of management is extremely thin. Gundemark, for example, had a staff of three to help him run this $300 million global relays business. And, in addition to his responsibility as global BA head, he also ran the Swedish relays company!

At the top of the company is the Group Executive Management, which, at the time of our research, consisted of seven managers besides Barnevik. Three of the seven headed the major regions in which the geographic spread of the company is concentrated— North America, Europe, and the Asia Pacific. For example, Eberhard Von Koerber was responsible for the European region, while Gerhardt Schulmeyer was his counterpart for North America. The other four executives in this group each head one of the four "sectors" in which ABB's more than fifty Business Areas are grouped: Goran Lindahl, for example, was the head of the Power Transmission and Distribution sector.

This was the group that Percy Barnevik chaired, and collectively it provided overall direction and leadership from its corporate headquarters in Zurich. Yet, under the rigorously enforced 90 percent rule, the corporate staff was trimmed from more than 2,000 to only 150, with most human, technological, and financial resources having been transferred to the operating companies.

How does such an organization function? How can you run a large global company, with a broad portfolio of technology-intensive and mutually interdependent businesses, with 1,100 highly autonomous companies, a central staff of only 150, and only one intermediate level of management? To understand that, it is necessary to look behind this structural framework to see the work connections between the different managers. In other words, we have to supplement the two-dimensional photograph of the organization represented in Figure 7.1 with an X-ray of its management process, represented in Figure 7.2, and eventually with a CAT scan image that allows us to observe the dynamism of those processes.

Figure 7.2
ABB: The Management Process

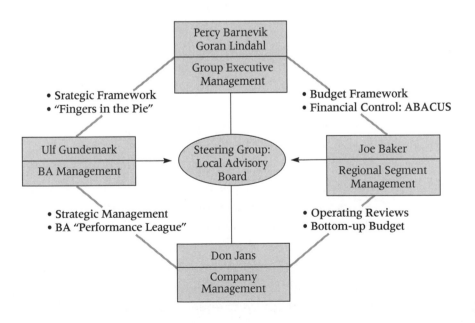

An X-ray of ABB's Organization

In the conventional image of hierarchical companies, power and authority accumulate at the top. Even in a dramatically delayered and radically decentralized structure like ABB's, the assumption is that the corporate-level group executives drive the company through their control over its strategy. While this is true in part, it does not provide an accurate view of how ABB's top management views its task. "Deciding on strategy is only 5 percent to 10 percent of the job," said Barnevik. "The other 90 percent to 95 percent is making it happen."

So while Barnevik has clearly been deeply involved in leading the company's key strategic commitments—to stay focused in electrotechnical products, for example, to execute a major shift of resources from Western Europe to North America and Asia, and to bet on the growth of developing markets like India and China—the analysis and decision making involved in reaching such conclusions have been far less demanding than the huge time commitment Barnevik and his group executives have made to the task of embedding broad corporate purpose and shaping a set of values to guide behavior. Traveling over two hundred days a year, he estimates that he meets personally with over five thousand of the company's managers annually.

While outlining his broad strategic vision, Barnevik is also likely to probe managers for ideas about what the company could be doing to help their units become more competitive. Using some of the two thousand overhead transparencies he travels with, Barnevik does much to ensure that the organization understands and is aligned around his vision. Equally important, he emphasizes the "how to"—the management policies and principles that are vital for effective implementation. For example, he might expound on his 7–3 formula ("It's better to decide quickly and be right seven out of ten times than postpone. The only unacceptable action is to do nothing"), and he almost certainly will preach a doctrine of individual accountability. All of his efforts are focused

on giving frontline managers the vision, the resources, the impetus, and the encouragement to feel a sense of ownership and the
freedom to exercise it. In short, Barnevik wants to counterbalance
the top-down bias of the hierarchical structure with a powerful
sense of initiative and drive that comes from the bottom up.

At the middle levels of the organization, the actions of managers like Gundemark and Baker are far from control-dominated.
While each of these managers has a primary responsibility for
achieving results—Gundemark's BA dimension focuses on global
strategy and Baker's regional segment responsibility emphasizes
financial performance—they achieve them through nontradi-
tional methods. Most key decisions for each of the operating
companies are hammered out in the steering committee that has
been constructed for each frontline company. These are minia-
ture boards of directors, and in addition to Baker and
Gundemark, the membership of Jans's steering committee also
included a couple of other frontline company presidents and
some technical and financial experts who brought perspectives
and experience to bear from other parts of the company. Through
the effective use of this forum, both the operating budget and
strategic-planning processes became constructive dialogues
rather than imposed objectives. In addition, the usual matrix
problem of frontline managers being squeezed by their bosses'
conflicting interests was avoided.

Senior managers at ABB also created other integrative forums
and made them important parts of the management process. For
example, Gundemark exercised his responsibility for global strategy and cross-market policies in relays by creating a BA board
consisting of himself, his three-person staff of finance, technical,
and business development directors, and the presidents of the four
major relays operating companies, including Don Jans. And at the
operating level, he formed functional councils that linked R&D,
purchasing, and quality managers with their counterparts in other
companies, rotating their quarterly meetings as a way of capturing
and transferring best practices. This benchmarking was reinforced

by an internalized control mechanism that Gundemark called his "performance league." By distributing the ranking of all relays companies on key performance criteria, he found that managers in lower quartile companies quickly linked up with those near the top to learn how to improve their performance in inventory management, quality level, and the like.

If the thrust of top-level management's actions was to offset top-down processes with bottom-up initiatives, then the impact of those in the middle was to supplement traditional formal vertical linkages with new horizontal ones. With so few staff members to engage in traditional control activities, executives like Gundemark were forced to create shared supervisory forums like the BA board, the steering committees and functional councils, and self-regulating mechanisms like the performance league.

Finally, the static nature of the formal structure is given life by an information system and a management style that makes this process "X-ray" of ABB more like a CAT scan in which the internal functions can be seen in dynamic interaction. The company's PC-based ABACUS financial reporting system was created not only to provide uniform, fine-grained measurements of all key dimensions of the company's operations; it was designed to democratize information by making reports available in the same format and at the same time to everyone throughout the company. The objective was first to serve the needs of operating-level managers in identifying and diagnosing problems, and secondarily to provide senior management with a means of monitoring performance. With a style that Lindahl described as "fingers in the pie" management, those at the top never hesitated to reach down to the front lines if they sensed a problem. But the objective was to help rather than interfere. Recall the three questions that Lindahl always asked: "What's the problem? What are you doing to fix it? How can we help you?"

The planning process was equally unconventional. Lindahl saw his key role not as aligning, integrating, and blessing the plans of his various BA and regional managers in a ritualized annual review, but of questioning, probing, and challenging them in bimonthly meetings. He developed scenario exercises to force them to think about how they might change their strategic postures or priorities in response to various unplanned political, economic, or competitive developments.

Only by studying ABB's internal processes did we begin to get a real sense of how the company works. The top-down bias of the hierarchy has been overlaid with a strong bottom-up drive and initiative; the vertical, financially dominated communications have been supplemented with horizontal knowledge-intensive interactions; and the static tasks and responsibilities searching for the equilibrium of fit and alignment have been replaced by more flexible roles and relationships operating in dynamic disequilibrium.

THE COMPANY AS A PORTFOLIO OF PROCESSES

Our interest in ABB did not arise from its visible strategic moves or even its highly visible CEO. It was triggered by the transformation of Don Jans. What is it about the overall organization of ABB that led a veteran manager like Don Jans "to rediscover management"? How can we understand the difference in the organization of a company like ABB beyond its matrix structure or its ABACUS systems? After examining the operations of companies like ABB, Canon, and 3M, we concluded that we could *not* understand organization in terms of the very different, and often quite ungainly formal structures of these companies. Instead, we began to conceive of them as a portfolio of processes, and through that lens we could see some remarkable commonalties among these three companies and others that were in the process of becoming Individualized Corporations.

It should be clear that when we say process, we do not mean either the day-to-day operational processes like order entry or inventory control that have been the focus of process reengineering, or even the strategic processes like new product development or the integrated logistics chain. What we are referring to are the processes that operate one level above—the core organizational processes that overlay and often dominate the vertical, authority-based processes of the hierarchical structure.

At the heart of ABB—and at the heart of 3M, Canon, and ISS as well—lie three such core organizational processes. The first process—which we call the *entrepreneurial process*—produces and supports the opportunity-seeking, externally focused entrepreneurship of frontline managers. The *integration process* overlays the advantages of bigness—of size, scale, and diversity—on the advantages of smallness—flexibility, responsiveness, creativity—by linking the dispersed resources, competencies, and businesses of the company. And the third, the *renewal process,* creates and sustains its capacity to continuously challenge its own beliefs and practices, thereby revitalizing the strategies that drive the businesses. By creating and managing these processes a company is able to develop and sustain the three characteristics of the Individualized Corporation described in part 1 of this book.

The Entrepreneurial Process

The divisionalized hierarchy has been pilloried for killing entrepreneurship in large corporations. But, as the cases of 3M, Canon, and ABB confirm, it is not structure per se that is to blame. What has really caused this erosion are the assumptions about the roles and tasks of different management groups that were implied by that structure.

Over the last five decades, even though the specific structures of companies have evolved from Alfred Sloan's original design, in the absence of clear and well-articulated alternatives, managers have continued to play their historic roles. Top-level managers have continued to act as chief corporate entrepreneurs, setting corporate strategy and implementing it through their control over the resource allocation process. Middle managers have remained focused on the proper fulfillment of the demands of checks and balances, playing the role of administrative controllers. And swamped by direction and control from above, frontline managers have resigned themselves to the role of operational implementers, responding to the demands of internal organizational processes rather than focusing on external opportunities.

What is different about ABB is that it has fundamentally changed these assumptions about these roles and in doing so has created an entrepreneurial engine deep inside its organization. This carefully executed change should be clearly distinguished from the "quick fixes" of skunk works, internal venturing, or "interpreneurship" that have risen and fallen as management fads. The change within ABB is a transformational change, modifying old assumptions about responsibility and accountability at all levels of management in a way that builds a long-term capability.

At the same time, however, it is important to underline what ABB is not: It is not an organization of frontline cowboys held together by senior management acting as a venture fund team. Instead, it has redefined roles and relationships at all levels of management, to drive the company's ability to constantly seek and exploit new opportunities. It is this companywide process that brings the large-company advantages to the frontline managers and distinguishes the Individualized Corporation from those companies that have tried to achieve the same results through faddish quick fixes like internal venturing and "intrapreneurship" programs.

The entrepreneurial process is based on frontline, middle, and top-line managers playing very different roles than they have been used to. Frontline managers—like the heads of ABB's 1,100 companies and 3M's 3,000 business units and project teams—must evolve from the traditional role of implementer of top-down decisions to become the primary initiator of entrepreneurial action, creating and pursuing new opportunities for the company. Middle managers like Baker and Gundemark are no longer preoccupied with their historic control role, but have instead become a key resource to the frontline managers, coaching and supporting them in their activities. And top management, having radically decentralized resources and delegated responsibilities, focuses much more on driving the entrepreneurial process by developing a broad set of objectives and establishing stretched performance standards that the frontline initiatives must meet (Figure 7.3).

Figure 7.3
The Entrepreneurial Process

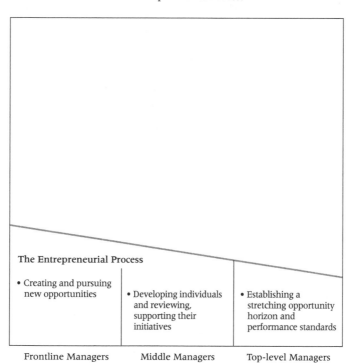

The Entrepreneurial Process

• Creating and pursuing new opportunities

    • Developing individuals and reviewing, supporting their initiatives

        • Establishing a stretching opportunity horizon and performance standards

| Frontline Managers | Middle Managers | Top-level Managers |

The fundamental difference between the ABB organization and the classic hierarchical company is brought sharply into focus with the changes that occur in organizational processes and management roles every time ABB acquires a company operating in the more traditional mode. With its acquisition of Westinghouse's North American power transmission and distribution business, for example, ABB had to overlay its radically decentralized structure and its philosophy of frontline entrepreneurship on a much more hierarchically controlled management system. The impact on the management of the relays division was typical of the changes occurring across the whole business.

Because of Westinghouse top management's unwillingness to support investments in what was seen as a mature business, Don Jans—then heading the relays division for Westinghouse—had been unable to convince his bosses to invest to any significant degree in solid-state and microprocessor technologies, despite the fact that his aging electromechanical product line was coming under increasing competitive threat. Under ABB's ownership, however, management roles and relationships changed radically. In his newly independent little relays company, Jans capitalized on his new strategic freedom and financial flexibility to start a small-scale effort to develop microprocessor-based products. As the initial experiments yielded some positive results, the scope of the activity was gradually broadened and the company was able to develop innovative new solid-state products in time to head off an aggressive new competitor who had begun to take advantage of the product-market gap.

Although Jans was certainly aware of ABB's high standards and performance expectations, the tyranny of a control-based hierarchy above him was removed. As head of ABB's worldwide Relays Business Area (BA), Gundemark saw his role as supporting Jans's initiative on microprocessor-based products, incorporating it into the broader global strategy to lead this industrywide technological conversion. As a member of the steering committee that acted as a board for Jans's local company, Gundemark was able to ensure that the proposal was supported and funded within weeks rather than the months, or even years, the formal capital budget request would have taken in Westinghouse—and even then would probably have been denied. Beyond funding approval, Gundemark provided Jans with advice and support in pursuing a strategy that involved a risk of short-term profit decline. For example, when a U.S. revenue shortfall led Jans's geographic boss, Joe Baker, to push for cuts in the $1.5 million budget Jans had obtained for microprocessor product development, Gundemark arranged to provide technical support from the Swedish relays company.

At the corporate level, Lindahl worked at building a shared commitment with his management team "to conquer the globe in power transmission." Expecting his BA heads and company managers to take the leadership in developing their business strategies, he saw his role as questioning and challenging. Beyond his strategy-testing scenario exercises, Lindahl also lived by the motto "What gets measured gets done." Having set high standards and ensured his management was working toward stretch targets, Lindahl made sure he did not slip into a mode he described as "abstract management" characterized by distant and impersonal control of large agglomerated organizational units through sophisticated but remote systems. He ensured the broad mission and the specific objectives were being achieved by practicing "fingers-in-the-pie management," a style that allowed him to contact Jans or other company managers directly to inquire about their business, encourage their initiatives, or offer help when performance was slipping off track.

ABB's remarkable success in transforming tired old-line companies into entrepreneurial competitors has caused the company to become an organizational benchmark that many others have tried to emulate. Some have copied the global matrix structure in the belief that it holds the secret to being able to internalize the complexities and contradictions of their operating environment; others have modeled their management systems after ABACUS, diffusing detailed but consistent information deep into the organization; and more still have been impressed enough by the philosophy of radical decentralization that they too have created independent frontline operations in which they hope entrepreneurship will flourish.

The problem is that while these companies copied the photograph, they ignored the lessons of the X-ray. The key to creating entrepreneurship is to build an entrepreneurial process by redefining management roles and relationships. The "new" definitions are, in many ways, antithetical to old tasks and responsibilities prescribed by the formal hierarchical structure. Yet at the base of this is an abiding faith in people that allows for true empowerment.

The Integration Process

In a world of converging technologies, permeable industry boundaries, and interdependent global markets, the entrepreneurial process alone is not sufficient to maintain competitive viability; the Individualized Corporation also requires a strong integration process to link the company's diverse assets and resources into broad corporate capabilities and to leverage those capabilities to create distinctive advantages that support existing businesses and help the company enter new businesses and markets. In the absence of such an integration process, decentralized entrepreneurship may lead to some temporary performance improvement, but long-term development of new capabilities and businesses will be seriously impaired.

Just as frontline managers like Jans play the pivotal role in the entrepreneurial process, it is those in the middle, like Baker and Gundemark, who play the anchor role in the integration process. However, just as the entrepreneurial process requires the complementary contributions of middle and top-level managers, so does the integration process require intensive involvement and support at all the levels of the organization (Figure 7.4).

At the top level of ABB, for example, the group executive supports the middle managers' pivotal horizontal linkage role by devoting enormous effort to creating a sense of shared organizational identity—what Barnevik calls the "glue"—and by building organizational norms that value collaboration—the "lubricant" in Barnevik's terminology. While the former provides the focus to bind the disparate efforts of the frontline entrepreneurs, the latter fosters the linkages that intensive knowledge transfers require. Without such an organizational context, the centrifugal forces driving independent entrepreneurial units can quickly result in fragmentation, isolation, and interunit competitiveness to create barriers and defenses against internal flow of knowledge and expertise.

ABB's statement of its values in the company's "policy bible" defines clearly the expectation that individuals and groups interact "with mutual confidence, respect and trust . . . and remain flexible, open and generous." Barnevik and his colleagues in the Executive Committee see it as their primary responsibility to translate those values into action through their appointment of individuals reflecting such behaviors to key positions, their imposition of sanctions against those violating the values, and through their own role-modeling behavior.

Figure 7.4
The Integration Process

The Integration Process		
• Attracting and developing competencies and managing operational interdependencies	• Linking dispersed knowledge, skills, and best practices across units	• Institutionalizing a set of norms and values to support cooperation and trust
The Entrepreneurial Process		
• Creating and pursuing new opportunities	• Developing individuals and reviewing, supporting their initiatives	• Establishing a stretching opportunity horizon and performance standards
Frontline Managers	Middle Managers	Top-level Managers

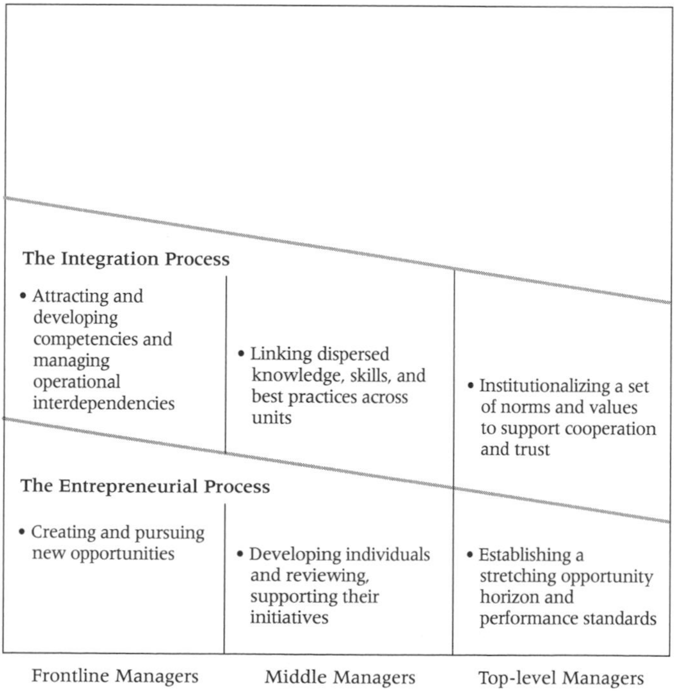

These strongly ingrained corporate norms of mutual trust and support have created an environment that encourages frontline managers to reach beyond the bounds of their own formal responsibilities and rewards them for doing so. Top management recognizes and rewards those who are seen as "givers" in ABB terminology—managers able to attract and develop talented people who become internal candidates for other parts of the organization. And ABB's top leaders also explicitly require managers to be "effective team players—people who can bridge contradictions, build mutual support, and achieve consensus," as the policy bible puts it.

By encouraging such behavior among frontline managers, the company has created an environment in which many of the cross-unit conflicts and operating interdependencies are resolved without intervention. In the relays business, for example, problems caused by ABB companies competing against each other in export markets were resolved through negotiations among the relevant marketing managers rather than by senior management. Indeed, when the marketing managers asked for an arbitrated solution, senior management pushed the issues back down, insisting on a unanimous negotiated agreement.

However, ABB and companies like it cannot rely on "spontaneous combustion" to drive the intensive knowledge sharing that is required if they are to develop organizational learning as a source of competitive advantage. While the top-level context setting and the frontline personal networks can provide the enabling conditions for this vital horizontal process, it is the middle managers who are the best placed to encourage the cross-unit linkages. Yet, historically, it has been at this level of the organization that the highest barriers to horizontal transfer of knowledge and expertise have existed, largely due to the management philosophy behind the divisionalized organization structure. By fragmenting the company's functional capabilities, while simultaneously creating a requirement for division managers to maximize the performance of their division, the old model impeded the reintegration of the compartmentalized resources and capabilities.

Managers in almost every company we have seen recognize the need to integrate their corporate resources and to link and leverage the knowledge and expertise of their people and organizational units. Indeed, the need is manifest in the excitement generated since the late 1980s by the concept of managing a company not as a portfolio of product-market positions but as a bundle of "core competencies." There are few companies today that have not invested a significant amount of management time and money trying to identify, develop, and exploit their own core competencies.

Unfortunately, in the implementation—if not in the intent—of this powerful concept, the desire to create and use core competencies often deteriorated into just another version of top-down corporate control. Particularly at corporate headquarters, functional staffs found a solution to the problem of their increasing marginalization as power was decentralized to business units and the headquarters groups were being destaffed. Latching on to the word *core*, they interpreted the notion to mean that corporate executives must define, develop, and control the company's strategic competencies, making them available to operating units in the same way they allocated capital and other resources.

The integration process in companies like ABB defines a model that is better described as managing by distributed but integrated competencies. In this model, senior management's role is viewed not as that of defining, controlling, or allocating competencies but as creating an environment that ensures that competencies are developed deep within the organization and building the horizontal linkages that allow distributed competencies to be integrated and leveraged as broad organizational capabilities. This is what Gundemark was doing by creating functional councils designed to transfer best practices developed by leading-edge companies, and by creating steering committees structured to provide frontline managers with advice and support from their colleagues. In other words, the purpose of the integration process is to ensure that the learning organization overlays and supplements the entrepreneurial organization instead of supplanting it.

The Renewal Process

Although the traditional divisional structure can create a highly efficient implementation machine that allows a company to refine its operations continuously, it has been less effective in renewing them. Based primarily on vertical processes of information processing that convert data into information and information into knowledge, it lacks any mechanism for challenging existing beliefs and strategies. As a result, unquestioned and unquestionable verities from the past have often become enshrined as "the company way."

The renewal process is designed to counteract this trend. Its objective is to challenge a company's strategies and the assumptions behind them. In that sense, while the entrepreneurial and integration processes support present strategies, the renewal process aims to disrupt them. And, like the other two processes, it too is defined by a set of top, middle, and frontline management roles that not only drive the process but also embed it in the organization's ongoing activities (Figure 7.5). While the frontline managers are the key drivers of the entrepreneurial process and middle management provides the anchor for the integration process, it is the top management of a company that takes the lead in inspiring and energizing the renewal process.

It is in this context that top managers have the crucial responsibility of serving as the source of organizational disequilibrium. Yoshiro Maruta of Kao Corporation, profiled in chapter 6, is a perfect example. By creating an exciting sense of purpose and ambition, he was able to create enormous stretch beyond the organization's existing performance level, and indeed, beyond its existing resources and capabilities.

Although his management style is quite different from Maruta's, ABB's Percy Barnevik shares much of his Japanese

Figure 7.5
The Renewal Process

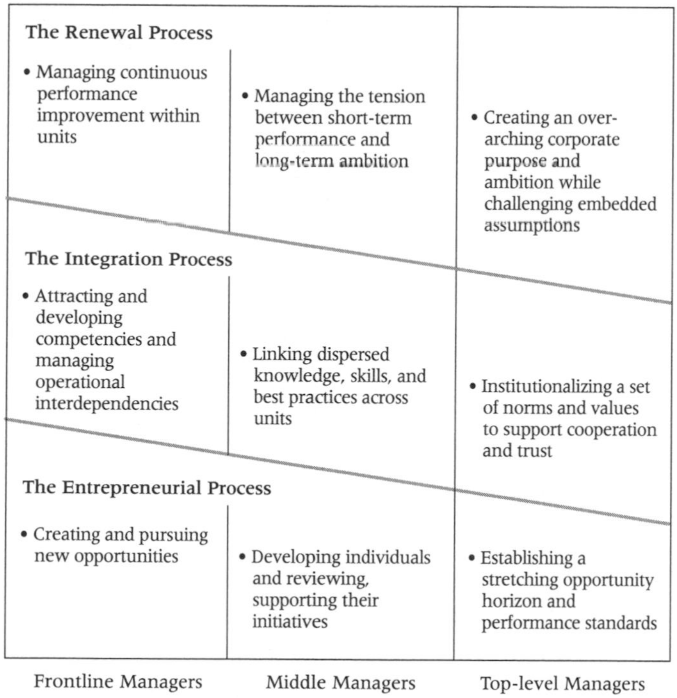

The Renewal Process		
• Managing continuous performance improvement within units	• Managing the tension between short-term performance and long-term ambition	• Creating an over-arching corporate purpose and ambition while challenging embedded assumptions
The Integration Process		
• Attracting and developing competencies and managing operational interdependencies	• Linking dispersed knowledge, skills, and best practices across units	• Institutionalizing a set of norms and values to support cooperation and trust
The Entrepreneurial Process		
• Creating and pursuing new opportunities	• Developing individuals and reviewing, supporting their initiatives	• Establishing a stretching opportunity horizon and performance standards
Frontline Managers	Middle Managers	Top-level Managers

counterpart's basic philosophy. Unlike many of his contemporaries, Barnevik does not accept the logic of incrementalism whereby the organization's aspirations are determined by the goals of the individual managers according to their perceptions of the past. Instead, he has focused ABB on a very stretched, future-oriented corporate mission that is communicated with conviction and intensity so as to inspire the simultaneous raising and convergence of individual aspiration levels within the company.

While many companies have aspired to the voguish objective of "creating a shared vision," Barnevik seems to have moved well beyond the rhetoric toward making this an operating reality at ABB. He has done so through several means. The company's mission statement articulates a broad corporate purpose with which all organization members around the world can identify: "to contribute to environmentally sound sustainable growth and make improved living standards a reality for all nations around the world." More than a statement of product-market strategy, this is an organizational purpose that can provide individual employees with a pride in committing their energies to its achievement. However, the mission statement then expands that altruistic purpose, tying its achievement into a more focused managerial objective: "to increase the value of our products based on continuous technological innovation and on the competence and motivation of our employees . . . becoming a global leader—the most competitive, competent, technologically advanced and quality minded electrical engineering company in our fields of activity." This translation of the broad mission to a strategic objective lends it a sense of organizational reality and legitimacy, and gives it more managerial power and relevance. Barnevik further operationalized the broad vision by expressing the goals in financial performance terms: operating profit at 10 percent of sales and a 25 percent return on capital employed by the mid-1990s.

Beyond the indirect role of challenging the organization by stretching its sense of purpose and ambition, top management often has to play a more direct role in driving the renewal process. However much an organization is conscious of the need to question existing dogma and established methods, the most far-reaching strategic challenges to drive corporate renewal often need to be initiated from the top. For example, recognizing that ABB was too dependent on a stagnant European market, Barnevik set an objective to expand its North American sales to represent 25 percent of the company's total. To jump-start the massive change, he acquired Combustion Engineering and Westinghouse's power transmission and distribution business in the United States, grafting these operations on to ABB's existing BAs. Not only did such a move boost the North American share of ABB's sales from 12 percent to 18 percent, it also exposed the European-dominated business to North American technical developments (such as process control automation) and management practices (such as time-based management) that had not previously been well understood or effectively practiced in the ABB system.

The dual impact of a clearly defined anchor of corporate purpose juxtaposed against a slightly dissonant corporate ambition is felt particularly strongly in the frontline units. It is here that managers confront the gap between current performance and the aspiration level defined by the visionary purpose statement and the stretching ambition. If this process is to remain in balance, however, this tension created in the frontline units must be resolved. And without a high level of credibility and trust, such resolution can easily degenerate into horse-trading or grand compromises. Creating and maintaining credibility and trust in the system, therefore, becomes a key requirement and, in fact, defines the main task of middle management in the renewal process.

One important means of carrying out this role is through the stewardship of the various horizontal coordination mechanisms we have described. All the boards, committees, and project teams that Ulf Gundemark created and managed served not only as channels for the communication and transfer of knowledge but were also explicitly designated as forums where managers could negotiate differences and resolve conflicts in an open and legitimate manner.

Middle managers also ensure the legitimacy and credibility of this tension-filled process by creating a decision-making context that is both participative and transparent. To ensure the credibility of top management's ambitious objectives, for example, Ulf Gundemark formed a Relays Vision 2000 Task Force of nine operating managers from various frontline companies and charged them with the task of translating Barnevik's broad vision of ABB's place in the global power equipment industry into specific strategies for the relays businesses. After six months of intensive effort this task force developed a set of self-funding proposals that ranged from strategic investments aimed at expanding into new products (such as metering) and new markets (such as telecommunications) to plans for increasing employee motivation and organizational effectiveness. With a legitimacy and credibility that came from being defined by peers rather than being imposed by the top management, the strategy was accepted by frontline managers such as Jans who began trying to meet the ambitious challenges by expanding beyond their existing business boundaries into the new products and markets legitimized by Vision 2000.

A NEW ORGANIZATIONAL MODEL

Defining ABB in terms of its three core organizational processes is a very different way to think about an organization. The more typical focus on redrawing the boxes and lines of the formal organization charts reinforces an engineering approach to organization building—an approach based on a systematic disaggregation and reaggregation of the company's tasks and responsibilities. Instead, by thinking in process terms, we can envision the ABB organization in terms of the roles played by the different managers, and the relationships among them through which the work gets done. This view focuses attention on the proposition that an organization is more than just an economic entity composed of a hierarchy of tasks and responsibilities. Above all else, it is a social institution, made up of people and defined by their roles and relationships. That view captures the essence of how Barnevik's approach to the organizing task differs from that of most chief executives, and how the Individualized Corporation differs in its fundamental philosophy from that of the traditional divisionalized organization.

It is worth reemphasizing that the essence of this distinction does not lie in the formal structure. ABB can be "fit" into the traditional organizational model shown on the left-hand side of Figure 7.6. The darkened small circle represents Jans, while the other circles represent his colleagues at the head of ABB's 1,100 local companies. Jans's dual reporting relationship to Baker and Gundemark can be represented, as can the chain of command through Lindahl to Barnevik at the apex of the hierarchy. Such a representation of ABB is essentially accurate—it illustrates exactly the formal responsibility-based relationships among these individuals.

Figure 7.6
A New Organizational Model

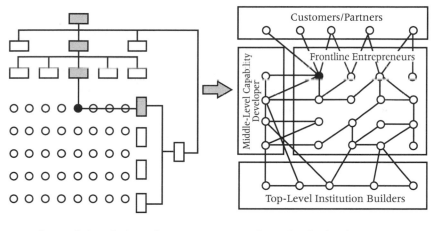

The Traditional Hierarchy   The Individualized Corporation

However, if the purpose of an organization chart is to describe how the company actually functions and who adds value in what way, then this representation of ABB is wrong. The company's organizational philosophy is far better represented by the figure on the right. Instead of emphasizing managers' positions in a hierarchy—top, middle, or frontline—this model places more emphasis on their roles.

To sum up:

• Frontline managers, heading small, disaggregated, and interdependent units focused on specific opportunities, are the company's entrepreneurs. They are the builders of the company's businesses and competencies and take full responsibility for both the short-term and the long-term performances of their units. Importantly, they face out to an external environment with which they build strong contacts and relationships, rather than upward into a hierarchy from which they expect direction and control.

- Like coaches who leverage the strengths of individual players to build a winning team, middle-level managers link these separate businesses and leverage the resources and capabilities developed in each of them. Overall, they play the role of capability developers—developing both the skills and competencies of the individual frontline managers through mentoring and guidance, and also the overall capabilities of the organization by integrating the diverse capabilities of the frontline units across businesses, functions, and countries.

- Top management provides the foundation for this activity by infusing the company with an energizing purpose—a sense of ambition, a set of values, an overall identity—so as to develop it as an institution that can outlive its existing operations, opportunities, and executives. Like social leaders, top management creates the challenge and commitment necessary to drive change and ensure that the company continuously renews itself. Rather than trying to control strategic content, top management focuses much more on shaping organizational context.

This chapter opened by describing the differences in the organizational structures of 3M, Canon, and ABB. With this process-based view of the organization we can now highlight their similarities. Take 3M. This is the company that pioneered the creation and institutionalization of entrepreneurial processes deep in the organization, and our earlier description of Andy Wong's optical systems business unit provided a classic example. Despite the differences in the structures in which they operated, Paul Guehler was managing a very similar integration role as head of 3M's Safety and Security Systems Division as Ulf Gundemark was leading in his role as Relays BA head for ABB. And like Percy Barnevik at ABB, 3M's "Desi" DeSimone focused most of his time and energy on creating an organizational context that supported continuous self-renewal.

At Canon, too, we witnessed the same configuration of roles and processes. Ryuzaburo Kaku launched Canon on its growth path by radically decentralizing the organization and by creating maximum leeway for each subsidiary to act on its own initiative. Describing Canon's organizational roles, he compared his own role to that of the "Shogun"—the head of the Tokugawa government that brought peace and prosperity to Japan by securing hegemony over the warlords and then granting them almost complete autonomy in their own territories. The unit heads were "daimyo," the warlords who were fully responsible for the success and the prosperity of their areas and people. "The only difference between Canon and the Tokugawa government," said Kaku, "is that while the regime was a zero-sum society, Canon's objective is to enhance the prosperity of all subsidiaries through efficient mutual collaboration."

This mutual collaboration is built into Canon primarily through the functional committees. The members of these committees lead the integration process, constantly pulling together teams and task forces, both within individual functions and across them, to create new technologies, capabilities, and businesses. A typical example was the role played by the head of Canon Development Systems (CDS), who created a cross-functional team from the optics technology group in the Canon Research Center, the camera manufacturing units in Canon Electronics, and Canon Sales, the company's marketing arm in Japan, to develop a new video camera. By incorporating all the relevant technical, manufacturing, and marketing expertise, the team was able to minimize the development risk as it brought this new product to market.

As the corporate leader, Kaku himself plays a role akin to that of Barnevik and DeSimone. Almost immediately after taking over as the president of the company, he launched the "premier company plan" that articulated the ambition of making Canon a premier company in Japan—a far cry from Canon's external reputation at that time. By 1990, as Canon had emerged as a highly respected company in Japan, Kaku raised the bar by challenging the organization to become a "premier global company." In operational terms, this required the company to earn profits in excess of 15 percent on sales, so that R&D expenditures could be increased by over 50 percent. Once again, the enormous sense of stretch this implied in terms of improving existing operations and building new competencies was captured by Kaku in rich imagery:

> By implementing our first plan for becoming a premier company, we have succeeded in attaining the allegorical top of Mount Fuji. Our next objective is Everest. With firm determination, we could have climbed Fuji wearing sandals. However, sandals are highly inappropriate for climbing Everest; it may cause our death.

If renewal is powered by the process of creative destruction, then no company we studied pursued it with as much vigor and courage as Canon. For example, Canon dominates the laser printer market, enjoying a staggering 84 percent global manufacturing share for this product. Yet, it is Canon that developed and aggressively pushed the bubblejet printer technology, which was designed to make laser printers obsolete. And, in emphasizing top management's role in driving this process, Kaku identified what is perhaps the single most important barrier to renewal in many companies:

> In order for a company to survive forever, it must have the courage to deny at one point what it has been doing in the past; the biological concept of "ecdysis"—casting off the skin to emerge to new form. It is difficult for human beings to deny and destroy what they have been building up. But if they cannot do that, it is certain the firm cannot survive forever. Speaking about myself, it is difficult to deny what I have done

in the past. So, when such time comes that I have to deny
the past, I inevitably would have to step down.

The Individualized Corporation can be examined in terms of its
core processes and the new management roles embedded within
these processes. Because these new roles are what lie at the heart
of the new organizational model, the key challenge in transforming a company into an Individualized Corporation lies in transforming the frontline, middle, and top-level managers so that
they are willing and able to play their respective roles of entrepreneurs, capability developers, and institutional leaders.

Chapter 8 will focus even more sharply on these roles, providing a rich elaboration of the value-added tasks and describing
how companies can recruit, develop, and deploy a new breed of
managers who have the requisite competencies to play these new
roles.

Developing Individual Competencies: Beyond the Russian Doll Model of Management

8

Although their recent performance may have caused the memory to fade, over the years, the Boston Celtics have won more National Basketball Association championships than any other team in the league. They achieved that record by building a highly effective organization that became the source of sustainable long-term competitive advantage. It was an organization characterized by the exceptional leadership ability of their general managers (as epitomized by the legendary Red Auerbach), the strong team development skills of coaches (such as Tom Heinson), and the outstanding on-court talent of players (like Larry Bird). But it is clear to everyone in the Celtics organization that the capable general manager, the savvy coach, and the star player all add value in very different ways. Although Auerbach's career demonstrates that a good player can occasionally evolve into a great coach, and even go on to become an exceptional general manager, such a progression is extremely rare. Success in one role is not an accurate predictor of performance in another. Heinson made the transition from player to coach with ease, yet was not seen to have general management potential. And despite the fact he was one of the game's most successful players ever, few expected Larry Bird to become a great coach like Heinson, let alone a general manager of Auerbach's standing.

Yet, when it comes to companies, most organizations operate on a long-standing and firmly established belief in what might be called a "Russian doll" model of management. Management has come to be structured as a neatly nested hierarchy of responsibilities built around a corporate system designed to focus all managers on the core task of bidding for resources, negotiating objectives, and meeting performance targets. At each level of the hierarchy, a manager is expected to gain the knowledge and expertise required to succeed at the next level of this overall planning, budgeting, and control process. Frontline managers who demonstrate skill in filling out the strategic-planning and operating-budgeting formats and in implementing their approved objectives are assumed to be ready to take on the middle-management role of administering the planning process and controlling its implementation; and the most effective of these administrative controllers form the pool from which the top-level managers are drawn to define the strategy, set the objectives, and allocate the resources.

Although the Russian doll model may have fit the assumptions on which divisionalized hierarchies like Westinghouse were built, in today's high-performance organization it is being swept away by the same transformational changes that are reconfiguring companies around a portfolio of processes rather than on a hierarchy of tasks. The new organizational model described in chapter 7 has profound implications not only for the roles and responsibilities of most management positions but also for the personal competencies and capabilities of those who fill the redefined roles.

These changes are causing a revolution in the whole area of human resource management, where companies must first develop a rich understanding of the different portfolio of attitudes, knowledge, and skills required for the frontline, senior, and top-level jobs—in effect, what differentiates a Larry Bird from a Tom Heinson from a Red Auerbach. Then, human resource professionals and corporate executives must revise their views of how to attract, develop, and deploy people who exhibit the desirable competencies in a way that turns the new organizational concept into an operating reality. In the process of making these changes, companies are creating a new management philosophy that is replacing the old Russian doll model that has constrained rather than enhanced companies' organizational capability.

NEW MANAGEMENT ROLES, NEW PERSONAL COMPETENCIES

One of the most powerful forces driving the transformational changes gripping so many companies worldwide is the shift that is occurring in the relative scarcity of resources. As knowledge begins to replace capital as the company's most valuable strategic asset, we have witnessed how the strategies, structures, and systems designed to distribute and control the use of financial assets are being replaced by corporate models more adept at developing and exploiting information and knowledge. But it is not just the organizational structures and processes that are affected. The new corporate model introduced at the end of chapter 7 is built on a set of management roles and relationships that are radically different from those defined by its predecessors. The first challenge for any company is to understand the nature of these new roles and relationships and the ways in which they change the nature of each manager's key tasks.

Tearing Up Old Job Descriptions

The Russian doll model of management was born and nurtured in the divisionalized hierarchy structure that was perfectly designed for the expansionary yet stable postwar era in which it flourished. With capital as the scarce resource, this organization was designed around management roles that were focused on the vital task of finding the most productive uses for companies' limited finances. The core responsibility of top management in such classic hierarchies as Westinghouse or Norton was to be the *strategic resource allocator,* evaluating the strategic plans and capital budgets generated deep in the organization and making choices among them. Middle management played the role of *administrative controller,* consolidating and filtering the plans and investment requests moving up the organization, and interpreting and controlling the strategic choices and operating objectives being set by top management. The primary function of frontline managers was to be the *operational implementers,* translating the directions and priorities they received from above into action and results.

If a company is to become an Individualized Corporation, its operating-level managers must evolve from their traditional role as frontline implementers to become *innovative entrepreneurs;* senior-level managers must convert themselves from administrative controllers to *developmental coaches;* and top-level executives are forced to see themselves less as their companies' strategic architects and more as their *institution builders.*

To flesh out these new roles, we will focus on a select group of managers in companies such as ISS, ABB, 3M, and McKinsey. Although they were clearly in transition when we studied them it was apparent to us that these firms were acting more under the emerging philosophy than the old Russian doll model. Yet because the transformation is not yet complete, and also because companies' strategic tasks and organizational capabilities vary so widely, it is important to view these companies as illustrations of the new framework of management tasks rather than definitive models.

The role of the frontline entrepreneur has been visible in the actions of managers like ABB's Don Jans, 3M's Andy Wong, and

ISS's Theo Buitendijk. In contrast to the classic operational implementers, these individuals added value not only by focusing on improving ongoing productivity in their frontline units but also by taking responsibility for continued growth through innovation.

Among the many tasks and responsibilities these and other frontline entrepreneurs undertook, three were central to the way in which their role was differentiated from that of the classic operational implementer (see Table 8.1). The most striking set of activities and achievements common to the frontline entrepreneurs were those related to their willingness and ability to create and pursue new opportunities. (Wong's dogged pursuit of the optical systems business comes to mind, as does Jans's innovative initiatives to turn around his struggling relays operation.) Furthermore, rather than playing the passive-dependent role defined by traditional capital-budgeting processes and manpower-planning systems, the corporate entrepreneurs saw their role as "doing more with less." (Wong's redeployment of technical resources to build a marketing capability is a classic example; so too is Jans's ability to develop a microprocessor-based product line by borrowing some key resources and leveraging others.) Finally, these entrepreneurial managers all exploited the greater freedom they enjoyed in their empowered organizations to ensure the continuous performance improvement of their frontline units. (The turnarounds achieved by both Wong and Jans are great examples, particularly the latter, where the size of expense cuts, working capital reductions, and profitability improvements were made more impressive by the fact that they were achieved *after* the tight controls for such objectives were removed.)

Similarly, senior-level executives at these companies exemplified value added as a result of their developmental capabilities rather than their control skills. Using their higher level position to provide support and coordination rather than as a source of control, these managers were able to bring the resources and experience of a larger company to bear on the

Table 8.1

Transformation of Management Roles and Tasks

	Operating-Level Managers	Senior-Level Managers	Top-Level Managers
Changing Role	• From operational implementers to aggressive entrepreneurs	• From administrative controllers to supportive coaches	• From resource allocators to institutional leaders
Primary Value Added	• Driving business performance by focusing on productivity, innovation, and growth within frontline units	• Providing the support and coordination to bring large-company advantage to the independent frontline units	• Creating and embedding a sense of direction, commitment, and challenge to people throughout the organization
Key Activities and Tasks	• Creating and pursuing new growth opportunities for the business	• Developing individuals and supporting their activities	• Challenging embedded assumptions while establishing a stretching opportunity horizon and performance standards
	• Attracting and developing resources and competencies	• Linking dispersed knowledge, skills, and best practices across units	• Institutionalizing a set of norms and values to support cooperation and trust
	• Managing continuous performance impovement within the unit	• Managing the tension between short-term performance and long-term ambition	• Creating an overarching corporate purpose and ambition

smaller operational units created to initiate entrepreneurial activity.

Three key differentiating tasks separated the senior managers of Individualized Corporations from their counterparts in classic divisionalized hierarchies. First, they spent a substantial amount of time and energy in developing the individual members of their organizations and in supporting their ideas and initiatives. Second, they leveraged the pockets of entrepreneurial initiative and capability they had helped to grow by linking dispersed resources and expertise and transferring best practices across units. And finally, they became the key means for helping the organization balance the inevitable tension between the pressure for short-term performance and the challenge of ambitious long-term visions.

And at the highest rung of Individualized Corporations, top-level leaders were adding value not so much through their abilities to develop great new strategic insights and drive them toward implementation but by engaging the organization with a clear yet broadly framed sense of direction, then gaining frontline and senior managers' commitment to giving it both meaning and energy. The top leaders' main contribution was to provide the organization with the vision and vitality to move beyond refining its past achievements to developing the ability to continuously renew itself.

The first of the three differentiating tasks at this level was the widespread tendency for the most effective leaders (Barnevik, Welch, and DeSimone, for example) to challenge conventional wisdom and established objectives, replacing them with a stretch in vision and higher standards. Second, this group was much more focused on embedding corporate values that supported cooperation and trust rather than competition and mutual suspicion. But above all, the distinguishing role of new leaders is their ability to create a sense of purpose and ambition that may give rise to a set of strategic objectives but are much more broadly defined.

Together, these new roles and the key tasks that defined them created a radically different model of management than could ever be envisioned through the old Russian doll perspective. Yet as companies developed their delayered, reengineered, and more empowered organizations and began redefining management roles at their core, they often discovered that the real impediment to realizing their new corporate vision was their existing personnel; some were simply unable to operate effectively in the new model.

But if defining the profiles of those likely to succeed in these new roles has been difficult, that is nothing compared to the challenge of finding managers with the required attitudes, knowledge, and skills to meet the profiles. As most companies soon discovered, there was little in their traditional human resource policies or practices that provided much guidance in meeting the new challenges.

The Siren Song of Leadership Competencies

Like nature, management abhors a vacuum, and in an environment in which companies are clearly searching for answers to the problems created by changing roles, a new cottage industry of competency profiling has emerged to solve them. Consultants are vying for companies' attention with a range of tools and philosophies, which they claim will help to identify, measure, and develop the leadership qualities companies need to manage the emerging corporate paradigm.

The impact of these experts has been widespread, and over the past few years leading companies worldwide have invested enormous amounts of time and money trying to define the ideal profile for their future leaders. Siemens, for example, has defined twenty-two management characteristics under five broader competencies of understanding, drive, trust, social competence, and an elusive quality it calls a "sixth sense." PepsiCo's ideal profile identifies eighteen attributes and groups them into the categories of how individuals see the world, how they think, and how they act. AT&T focuses on eight core dimensions of leadership, Philips defines twelve key competencies, and the World Bank has twenty vital personal characteristics.

Yet, despite consultants' prodigious efforts in designing questionnaires, conducting interviews, and running seminars, few of these programs have had the promised impact. There appear to be several reasons why. The first is the difficulty companies have had in integrating the leadership profiles into their existing personnel policies and practices. (In the words of one manager, "How do you recruit based on a profile that reads like the Boy Scout Code—trustworthy, loyal, thrifty, kind . . . ?") Equally problematic has been the sponsorship issue. Typically led by the human resources department on the basis of an outside consultant's analysis and recommendations, these leadership competency exercises often lack the broad-based internal credibility and frontline management support needed to redefine the qualifications of an entire management team.

The greatest drawback to these management competency exercises, however, may be that invariably they result in a single ideal profile. Such an assumption may not have been entirely irrational in an era where more symmetrical roles (typical of the traditional authority-based hierarchy) were the norm. But in a time in which delayered and empowered organizations are radically reshaping roles and relationships, such continued respect for the old nested Russian doll hierarchy model of management seems counterproductive. As organizations recognize the need for a more differentiated set of management tasks, they must also recognize the myth of the generic manager. In short, the Individualized Corporation not only embraces individual differences, it capitalizes on them in a differentiated set of roles that require different personal competencies.

NEW PERSONAL COMPETENCIES, NEW HR TASKS

As we studied companies in which managers were adapting to the new roles, it was easy to see the importance of defining the personal characteristics needed to succeed. But equally clear were some of the limitations of the competency profiles. Rather than basing them on surveys of current managers—a widespread consultants' approach doomed to define future leadership needs in terms of historic management competencies—we opted to observe firsthand those who had demonstrated their effectiveness in the new management roles. And rather than develop a list of generic competencies with universal application, we differentiated the competency profiles by basing them on managers who succeeded in adding value at each level of the posttransformational organization.

Yet despite the fact that these profiles were based on performance rather than opinion, the notion of individual competencies still seemed too vague and unfocused to be of great practical value. To earn the support of top managers, the concept had to be more sharply defined and better integrated with day-to-day human resource activities. In short, to be of practical value, competencies had to be linked to actionable decisions.

This recognition led us to develop a simple model that classified the broadly defined competencies identified for each role into three categories (Table 8.2). The first category includes intrinsic personal characteristics like the attitudes, traits, and values that are embedded in the individual's character and personality; the second category lists attributes like knowledge, experience, and understanding that generally can be acquired through training and career path development; and the third category features specialized skills and abilities that are directly linked to the job's specific task requirements and tend to be internally developed.

Through this categorization, we not only gave the concept of personal competencies a sharper definitional meaning but were also able to identify much more clearly how the different attributes of the profile were linked to various important human resource decisions. In particular, our observations led us to develop some hypotheses about how the framework could be used as a practical tool in selecting, developing, and supporting people in their new and sharply differentiated management roles.

Selecting for Embedded Attitudes

The high rate of failure among managers attempting to adapt to their newly defined tasks in restructured and reengineered organizations underscores the importance of identifying selection criteria to help predict success. Despite CEO Percy Barnevik's intense personal involvement in selecting candidates for the three hundred most senior management positions at ABB when the merged company was created in 1988, six years later over 40 percent of that group was no longer with the company. As Barnevik recognized at the time, the central problem was that few candidates possessed the personal competencies that matched the radically different organizational and managerial context he had defined for ABB.

The challenge of staffing a radically transformed organization places a huge strain on most organizations. Most companies in this situation select candidates for the new jobs primarily on the basis of the individual's accumulated knowledge and job experience. These are, after all, the most visible and stable qualifications in an otherwise tumultuous situation. Even more important for most companies, however, is the practical reality that selecting on the basis of knowledge and experience is a decision that can be made by default, simply by requiring the existing manager to take on the redefined job responsibilities.

But in the midst of a corporate transformation, past experience rarely proved to be a good predictor of future success. The most obvious difficulty was that much of the acquired organizational expertise was likely to reflect old management models and behavioral norms. But a far bigger problem was that the characteristics of those who had succeeded in the old organizational environment were actually antithetical to those required in the new one. As a result, many companies found themselves with a group of what Jack Welch called GE's "Type IV" managers—those who achieved results, but did so without buying into the new organization philosophy. In the end, Welch realized that he could never

Table 8.2

Management Competencies for New Roles

Role/ Task	Attitude/ Traits	Knowledge/ Experience	Skills/ Abilities
Operating-Level Entrepreneurs	*Results-Oriented Competitor*	*Detailed Operating Knowledge*	*Focuses Energy on Opportunities*
• Creating and pursuing opportunities	• Creative, intuitive	• Knowledge of the business's technical, competitive, and customer characteristics	• Ability to recognize potential and make commitments
• Attracting and utilizing scarce skills and resources	• Persuasive, engaging	• Knowledge of internal and external resources	• Ability to motivate and drive people
• Managing continuous performance improvement	• Competitive, persistent	• Detailed understanding of the business operations	• Ability to sustain organizational energy around demanding objectives
Senior-Management Developers	*People-Oriented Integrator*	*Broad Organizational Experience*	*Develops People and Relationships*
• Reviewing, developing, supporting individuals and their initiatives	• Supportive, patient	• Knowledge of people as individuals and understanding how to influence them	• Ability to delegate, develop, empower
• Linking dispersed knowledge, skills, and practices	• Integrative, flexible	• Understanding of the interpersonal dynamics among diverse groups	• Ability to develop relationships and build teams
• Managing the short-term and long-term pressures	• Perceptive, demanding	• Understanding the means–ends relationships linking short-term priorities and long-term goals	• Ability to reconcile differences while maintaining tension

Table 8.2 *(continued)*
Management Competencies for New Roles

Role/ Task	Attitude/ Traits	Knowledge/ Experience	Skills/ Abilities
Top-Level *Leaders*	*Institution-Minded* *Visionary*	*Understanding* *Company in Its Context*	*Balances Alignment* *and Challenge*
• Challenging embedded assumptions while setting stretching opportunity horizons and performance standards	• Challenging, stretching	• Grounded understanding of the company, its businesses and operations	• Ability to create an exciting, demanding work environment
• Building a context of cooperation and trust	• Open-minded, fair	• Understanding of the organization as a system of structures, processes, and cultures	• Ability to inspire confidence and belief in the institution and its management
• Creating an overarching sense of corporate purpose and ambition	• Insightful, inspiring	• Broad knowledge of different companies, industries, and societies	• Ability to combine conceptual insight with motivational challenges

bring about his massive corporate transformation until he replaced the Type IV managers—even those with the most knowledge and experience—with others whose behavior matched the new leadership profile.

Like GE, many companies are coming to believe that it is much more difficult to convince an authoritarian industry expert to become a coach than it is to graft industry expertise onto a strong people developer. It is a realization that is leading companies to the conclusion that an individual's characteristics should counterbalance and perhaps even outweigh acquired experience.

This change of perspective in selection criteria has not been as disruptive of existing personnel pools as many may have expected, however. To the surprise of leaders like Welch, Barnevik, and others, one of the unexpected benefits of organizational change is that it brings to the fore a group of talented individuals whose personal attitudes, values, and style fit the new requirements but had previously been hidden under the constraints of the old management model.

Another equally important and closely related conclusion companies arrived at: Just as management roles and tasks differed widely at each level of the organization, so too did the personal attitudes, traits, and values of those likely to succeed there. The universal "leadership competency profile" was of little help in identifying differences among individuals suited to the role of operating-level entrepreneur, senior-level developer, or top-management leader. This was certainly true at ISS, a previously nondescript Danish office-cleaning operation that has developed into a $2 billion global industrial-cleaning-services company. Among the most powerful forces propelling its growth has been the organization's awareness that it needed to recognize a different kind of individual to take on its key management roles at various organizational levels.

When Poul Andreassen became president in the mid-1960s, he undertook to inject new life into a demoralized organization, where management saw its key role as controlling unskilled employees to keep costs down. One of his first acts was to break the organization into profit centers; he then set about trying to find energetic, independent, and creative managers motivated to run these relatively autonomous units as if they owned them.

Such a profile fits Theo Buitendijk to a tee. A results-oriented competitor at heart, Buitendijk gave up his career in a heavyweight multinational oil company where he was frustrated by the constraints, controls, and lack of independence, to become the managing director of ISS's small Dutch cleaning business. In his typical creative and persuasive manner, Buitendijk had soon convinced his new team that they needed to expand beyond the highly competitive low-margin office-cleaning business and become a major force in the higher-margin segment of slaughterhouse cleaning. Despite his lack of industry experience or market knowledge, Buitendijk's engaging and competitive personality allowed him to make changes that doubled the reported sales of his frontline operation within two years.

However, Andreassen wisely judged that the personal profile of those able to move to the next level of management was quite different from that of the operating-level entrepreneur. Few were expected—or indeed had the ambition—to make that leap. One who did was Waldemar Schmidt, an operating-level entrepreneur who had turned around the company's Brazilian business.

Recognized as an individual who demanded a lot of himself and of others yet managed more by influence than authority, Schmidt was tapped to head up ISS's European division. Unconcerned by his total lack of knowledge of the European market or his inexperience in that very different part of the organization, Andreassen was more impressed by Schmidt's personal qualities and a management style that was reflected in the Brazilian unit's success in developing its people at all levels. (Indeed, the Brazilian Five Star training program was later adopted companywide, allowing cleaning supervisors to develop into unit managers through a five-stage program that provided training not only in cleaning systems and personnel management but also in customer relations, financial management, and leadership.) Schmidt's profile was ideally suited to Andreassen's view of the senior manager's key role as the developer and supporter of the budding frontline entrepreneurs, able to leverage their individual capabilities across the organization.

As expected, Schmidt became a champion of results-oriented competitors like Buitendijk whom he subtly coached and supported across what Andreassen called "the Chinese walls" that protected frontline operating units from excessive top-down interference. When the Dutch unit's profitability took a downturn as the company worked to develop its new abattoir-cleaning business, Schmidt offered his support and advice but did not intervene; when it succeeded, he celebrated the achievement by giving Buitendijk a platform at a senior management meeting at which to present his story; and, as others became interested, he made the Dutch unit ISS's center of expertise and gave it responsibility for supporting European expansion in the food sector.

While the creative, persuasive, and competitive Buitendijk possessed the personal qualities to succeed as an operating-level entrepreneur, and Schmidt's supportive, patient, and integrative style made him the ideal senior-level developer, an even more demanding set of personal characteristics were called for in an outstanding top-level organizational leader. When Paul Andreassen was named ISS's president, it was clear that he had been selected primarily for his innate personal qualities, since his knowledge of the industry, his experience in the organization, and even his leadership skills were all either limited or nonexistent.

As a young engineer who had been frustrated in a large, traditional corporation, Andreassen was much less interested in running operations than in building a more ambitious organization. He had a natural style that was both insightful and inspiring, and soon after joining the company he began articulating the view that ISS should no longer think of itself as just a cleaning company but as the world's best service organization. He wanted to make "ISS and service" analogous to "Xerox and photocopying."

Yet, beyond the grand unifying vision, Andreassen was also very much concerned with operations. Indeed, his most prominent personal characteristic was his willingness to subject everything to question and challenge. Even after thirty years in the company, he claimed that his best days were when he got out in the field "to stir things up." In the view of those who worked with him, it was this rare combination of insightfulness and inspiration, complemented by a challenging, questioning open-mindedness, that made Andreassen the institution-building visionary he became.

When Andreassen retired, the organization's first priority was to select an individual who, like him, could lead an organization that stretched and challenged its members while simultaneously making them loyal members of the institution. There are few individuals who, like Red Auerbach of the Boston Celtics, have the breadth of traits and the temperamental range to adapt to the varied roles and tasks of frontline entrepreneur, senior-level developmental coach, and top-level institutional leader, but Waldemar Schmidt proved to be such a manager. His breadth of vision, his openness and fairness, and his ability to stretch those around him made him a clear candidate for the top job.

One of management's most important challenges is to identify the personal characteristics that will allow an individual to succeed in a new, and often quite different, management role. And because there are so few like Schmidt, who have the predisposition to evolve from entrepreneur to coach to leader, it is equally important to recognize when someone who is successful at one level lacks the individual traits to succeed at the next. For those with the perceived potential, however, the next key challenge is to develop the knowledge and expertise that can support and leverage these desirable characteristics.

Developing People for Knowledge

In the wake of the transformation organizations have been undergoing, training and development have mushroomed into a massive undertaking involving huge investments as companies try to bridge the gap between human resource capabilities and the ideal leadership competencies. But there is only so much that traditional training activities can achieve, and as most high-performance organizations come to realize, this expensive management resource is most effective when it is focused on the task of developing competency elements that relate to an individual's existing knowledge and experience.

Poul Andreassen understood this well. And although he recognized that attitudes, traits, and values are changed only slowly and with difficulty through training, individuals selected for those attributes could quickly acquire the specialized industry or functional knowledge they needed to operate effectively in ISS. For this reason, he made training and development one of the hand-ful of functions he controlled directly at ISS's small corporate office. A classic example of how the company used this resource is provided by its Five Star training and development program, which took motivated and competitive frontline cleaning supervi-sors and, by systematically developing their technical and admin-istrative skills, grew them into operating-level entrepreneurs.

Most of the high-performance companies we studied made sub-stantial commitments to formal training. GE spent $500 million annually on training, and for decades its facilities at Crotonville have been turning out knowledgeable managers not only for its own prodigious internal needs but also for leading industrial com-panies, which regularly raid the company for its talent. In the con-sulting industry, Andersen Consulting invests an impressive 10 per-cent of its revenues in training. Its facility at St. Charles, Illinois, is the information technology equivalent of GE's Crotonville, and every one of its consultants goes through a thousand hours of for-mal sessions in the first five years. In these programs, Andersen focuses primarily on developing the knowledge base and expertise of its consultants, while acknowledging the valuable by-product of personal relationship building and corporate value sharing that also come out of such intense activity.

But no corporate training center or series of formal programs can develop employees to their maximum potential in an era where the relevant knowledge base for most jobs is changing rapidly and where, in any case, its application requires extensive hands-on experience. Indeed, GE acknowledges that despite its huge commit-ment to formal training, it contributes only about 10 percent of a manager's required knowledge and expertise development. In truly high-performance organizations, knowledge development is "built into the bloodstream of the organization's ongoing operations," as Dr. Yoshiro Maruta, CEO of Kao Corporation, puts it. Describing his organization as "an educational institution" rather than a corpora-tion, Maruta insists that "learning is a frame of mind, a daily mat-ter." Thanks largely to its institutionalized knowledge-building

capability to develop individuals, Kao has become universally recognized as one of the most consistently innovative companies in
Japan.

In the United States, a similar honor would undoubtedly be
held by 3M, another company with a strong commitment to
developing the knowledge and expertise of its employees. But a
company must know clearly what it is trying to develop. 3M's
model starts with the individual development objectives and
assessment process that every new employee must go through.
For example, a promising accounting clerk might be set the personal education goal of becoming a certified public accountant
within three years. That objective is supported through internal
business courses, company-sponsored participation in external
educational programs, and developmental assignments that provide experience in activities such as preparing financial statements and participating in audits.

Like Kao, however, 3M recognizes the limits of formal training
programs, and has built a major part of its knowledge development into the day-to-day operations of the organization.
Reflecting the oft-repeated mantra that "products belong to the
divisions but knowledge belongs to the company," 3M has created a work environment that not only facilitates the development of individual expertise and experience but also encourages
its effective transfer and application throughout the organization.
The way in which the company complements its formal training
programs with carefully managed on-the-job learning can best be
illustrated by examining the experience of managers at different
levels of the organization.

In 1989, Andy Wong, a young engineer in 3M's Optical Systems
(OS) unit, was asked to take on the leadership role of that struggling group. This quiet, even retiring young man had earlier been
identified by Ron Mitsch, a senior R&D executive, for his self-
motivation, tenacity, and creativity—all qualities that 3M looked
for in its frontline entrepreneurs. Wanting to give Wong the
opportunity to prove his potential, Mitsch systematically sought to

build the business knowledge and provide the functional breadth the young engineer would need to be an effective operating-level manager.

Recommended by Mitsch to head the OS unit's small technical development team, Wong exploited 3M's well-developed internal technical network to tap experts, attend seminars, and participate in forums to expand his knowledge of the various optical technologies the unit was pursuing. That knowledge, coupled with his innate insightfulness and his disciplined personal style, led him to make some key decisions that radically focused the team's previously fragmented activities. Building on his success in the OS unit's lab, Wong was then asked to take responsibility for the inefficient OS manufacturing operations. It was an assignment aimed at further developing his business knowledge and broadening his functional experience base while contributing to the company's needs. Despite his lack of prior production or logistical experience, Wong was able to access available in-house expertise to help him redesign the manufacturing process around the more tightly focused technologies, simplify the production process, and rationalize the logistics. The resulting 50 percent cost reduction was not only a tribute to Wong's learning, it was an affirmation that the well-established organizational norms of openness and mutual support were at work to ensure that "knowledge belongs to the company."

Through these developmental opportunities, Wong was able to broaden his knowledge of the business beyond his previous technology-based focus, and to expand his familiarity with the organization's resources beyond his old scientific contacts. This in turn gave him the personal credibility and operational effectiveness that prepared him for the role of OS unit general manager. Within three years, Wong had motivated his team and come up with a strategy that led to the successful development and launch of two major products, reversing more than two decades of declining performance and losses and saving the OS unit from almost certain shutdown.

But if frontline managers like Wong needed to develop their knowledge and expertise in the detailed operations of a particular business, senior-level managers like Wong's boss Paul Guehler had to broaden their understanding of key people, management practices, and organizational processes to allow them to succeed in a role that required a well-developed ability in personal coaching and organizational development. On this very difficult development path, formal training was less useful than broad organizational experience as a way of acquiring the required knowledge and understanding.

Like Wong, Guehler began his 3M career in the R&D lab, and was also identified as someone whose temperament led him to look beyond the technologies he was developing to the businesses they gave birth to. This budding entrepreneurial attitude led to his transfer to 3M's New Business Ventures Division where, like Wong, his natural drive and intuitiveness were harnessed to the task of exploring market opportunities and business applications for high-potential ideas and technologies.

Recognized for his people skills and his group-building abilities, Guehler was moved successively into different roles across three divisions, broadening his understanding of 3M's formal and informal processes and enriching his experience in managing them. In particular, his responsibility as business director for disposable products in Europe provided Guehler with valuable experience in managing complexity as he worked to integrate the activities of a diverse group of individuals and independent organizational units in highly competitive and fast-changing businesses.

By the time he was appointed general manager and later vice-president of the division to which Wong and his optical system unit belonged, Guehler brought not only hardheaded business knowledge but also a sensitivity to interpersonal management and organizational development—qualities that allowed him to play a key role in ensuring the success of Wong and his team. It was Guehler who set the struggling OS unit its demanding objectives and deadlines while at the same time providing it with cover from the growing number of voices calling for its shutdown; who required the team to work through a three-stage analysis and review process to convert their unfocused product dream into a viable business proposition; and who worked behind the scenes to secure the technical, marketing, and financial support the resource-starved unit needed to make the project a success. As Guehler saw it, his main job was "to create an environment where people can come forward with ideas and are supported to succeed . . . to develop the people who can develop the businesses." It was through his knowledge of 3M's people and processes, accumulated through his broad organizational development experiences, that he was able to do just that.

The knowledge and experience-based competencies required for the top management role are even more broadly defined and take even longer to develop. Indeed, ex–3M CEO Lou Lehr suggested that the top job at this company was most likely to go to someone only ten or fewer years from retirement for no other reason than it took thirty to thirty-five years to accumulate the breadth of experience to be effective as the head of this culturally rich, diversified company.

Livio "Desi" DeSimone, elected CEO in 1991, provides a textbook example of the care 3M takes in developing the knowledge and experience of those with the appropriate personal characteristics to succeed at the next level of management. Heavy in formal training early on and in experiential development later, DeSimone had moved up through technical, engineering, and manufacturing management jobs to assume general management roles, first as managing director of the Brazilian subsidiary and then as area vice-president of 3M's Latin American operations.

As a senior manager with the personal attributes of a potential top-level leader, DeSimone was exposed to the knowledge and experiences to prepare him for such a role. "There were always people taking an interest in my development," said DeSimone on assuming 3M's top job. To broaden his understanding of 3M's diverse businesses, markets, and technologies, he was assigned to head up each of the company's three sectors, one after the other through the 1980s. But the developmental assignment involved more than expanding DeSimone's business exposure. After spending most of his career focused on 3M's far-flung foreign operations, it was important that he spend time in Minneapolis–St. Paul to develop an equal understanding of the organizational context that framed and supported 3M's entrepreneurial innovation processes. This decade of corporate-level assignments immersed him in the structures, processes, and culture that were carefully nurtured at the organization's center. Finally, DeSimone's promotion to the 3M board in 1986 served multiple purposes. It was a reward to this effective manager, sending a clear signal that he was a serious candidate for the top job. It allowed him to bring his accumulated knowledge and insights into board deliberations. And, most important, it exposed DeSimone, who had been so focused on 3M, to the perspectives and experiences of directors who came from companies with different practices and approaches to management issues.

At 3M and at most other high-performing companies, training and development play a major role in building the different competency profiles required by the newly defined frontline, senior-level, and top management roles. But their approach is far from the traditional model built around carefully standardized training programs and a well-trodden career path of ticket-punching. Instead of trying to force employees into the one-dimensional mold of the "organization man," these companies use a portfolio of educational activities and career experiences to leverage very different natural traits and talents. As Andy Grove, CEO of Intel, put it:

> We originally started planning careers like Big Brother, but we gave it up. It was too complicated. Today, people in Intel move in every direction—upwards, sideways, downwards. It's a flexibility that allows people to develop themselves and find a place where they can contribute. . . . Careers here

advance not by moving up the organizational chart, but by filling the company's needs.

Coaching for Skills Mastery

Beyond an individual's personal attitudes and traits, and in addition to his or her developed knowledge and experience, there is a final set of personal competencies that we describe as skills and abilities. These are the attributes that are usually the best indicators of an individual's success, since they tend to be most directly linked to the key roles and core tasks that characterize a particular job. For an operating-level entrepreneur, for example, a key skill is often the ability to recognize and seize high-potential opportunities; in a senior-level development role, it is the ability to effectively empower those on the front lines that is vital; and for top-level leaders, the ability to create an exciting yet demanding work climate is a core skill that must be developed.

Even among those who seem ideally qualified in both personal qualities and by training and experience, many prove unequal to their new job. Some are simply unable to demonstrate the vital yet elusive skills that separate the merely satisfactory from the truly outstanding performers. The reason for much of this failure is that applied skills rely heavily on tacit understanding and capabilities that grow out of the interaction between an individual's embedded personal attitudes and traits and his or her accumulated knowledge and experience. The critical entrepreneurial skill of being able to recognize potential in business situations as well as in people is neither an innate trait nor is it easily trainable. However, it does appear to be a skill that can be nurtured and grown in individuals who are curious and intuitive by nature, and who have also developed a richly textured understanding of their particular business and organizational context.

Key to developing such skills is giving individuals the coaching and support that encourages them to apply their natural talents and accumulated experience to the particular challenges of the job. Indeed, this coaching role has become so important that it is now widely defined as a central management responsibility, particularly at the senior level of the organization.

We described how Andy Wong's success as a frontline entrepreneur was founded on his meticulous development—first as head of the OS unit's small laboratory, then as manager of its manufacturing operation, and finally as leader of the whole unit. But in the background was the vitally important coaching role played by Ron Mitsch, Wong's informal mentor, and Paul Guehler, his direct boss. Both helped Wong translate his natural talents and his rapidly expanding experience base into the much finer grained management skills he needed to turn around the stalled OS unit.

When talking about some of the important new roles managers are being asked to play—entrepreneur, for example, or leader— the widespread feeling among managers in high-performance companies is that the vital skills "can be learned but cannot be taught." Most companies in which personal mentoring is common believe that there is something almost mystical in the catalytic role that coaching plays in linking an individual's innate capabilities with his or her acquired knowledge and experience to allow such elusive skills to emerge.

One of the corporate leaders who believes most strongly in the power of personal mentoring is Roger Enrico, CEO of PepsiCo. Three years before he was named to the top job, Enrico stepped back from his operating responsibilities to devote half his time to the creation of personal links with a group of young executives he felt had the potential to become the next generation of presidents of PepsiCo's operating divisions. He organized a series of retreats and asked the carefully selected participants to bring a "big idea"—a proposal that they believed could have a major impact on their business.

Eschewing the usual outside management gurus and facilitators, Enrico conducted the sessions himself, meeting one-on-one with the participants to discuss their proposals in detail. After five days, the managers returned to their units to implement their projects. Ninety days later, the group reconvened to report on their progress and discuss follow-up action. Throughout the process and subsequently, Enrico remained on call as coach, adviser, and mentor to this group, mirroring his own relationship with Don Kendall, then CEO of PepsiCo—a supportive connection that had made such a strong impression on Enrico as a young manager.

One organization that has developed an extraordinary commitment to developing the skills and abilities of its people through intensive coaching is McKinsey. Although this was a deeply rooted tradition, there was a period in the early 1970s when the firm seemed to have temporarily lost touch with that commitment. Responding to a slowdown in its growth and, even more disturbingly, the perception that McKinsey's reputation as the consulting firm of choice to top management was fading, an internal Commission on Firm Aims and Goals bluntly concluded that McKinsey's "preoccupation with geographic expansion and new practice possibilities has caused us to neglect the development of our technical and professional skills."

This report led to the decision that the partnership must invest much more intensively in the development of its bright young recruits to become what the firm called "T-shaped consultants." Although specialized industry knowledge or functional expertise—the T's vertical spike—could largely be acquired through formal training and focused experience, it was the horizontal generalist problem-solving skills and client development capabilities that still distinguished McKinsey from its competitors. These were attributes that experienced consultants knew were best acquired through the intensive counseling and mentoring relationships that firm insiders often referred to as "the apprenticeship process."

The recommitment to the personal relationships that characterize an apprenticeship eventually led McKinsey to take the extraordinary step of redefining its core mission. To the long-standing commitment to serving the client that Marvin Bower had instilled into the firm, the partnership decided it should add a second goal—to build a great firm. Fred Gluck, McKinsey's managing director at the time, even suggested that this second dimension might become the dominant one:

> There are two ways to look at McKinsey. The most common way is that we are a client service firm whose primary purpose is to serve the companies seeking our help. That is legitimate, but I believe there is an even more powerful way for us to see ourselves. We should begin to view our primary purpose as building a great institution that becomes an engine for producing highly motivated, world class people who, in turn, will serve our clients extraordinarily well.

This focus on developing the skills and abilities of an exceptionally bright and highly motivated group of young recruits was clearly reflected in the career of Warwick Bray, a young Australian systems engineer who had earned an MBA before joining McKinsey's Melbourne office. During his first three years, he developed a real interest in the telecommunications industry and worked on several studies relating to the impact of deregulation on key companies. Besides developing this industry "spike," Bray was also becoming a more effective consultant, largely as a result of an intensive coaching process that began the moment he was assigned to his first engagement team.

It was firm practice for the engagement manager (EM) to sit down with each associate to discuss and agree on a personal development objective that the individual would work on during the study. In addition to the routine day-to-day coaching Bray received at the midpoint and end of each engagement, he also received detailed feedback and advice from the EM. Through this process, over the course of several studies, the young associate gradually developed his ability to recognize core issues, apply various problem-solving approaches, divide responsibilities, and integrate work—skills that were key to developing the creativity, insight, and initiative that characterized an effective frontline consultant.

Beyond this intensive on-the-job coaching, Bray received more continuous counsel and support from his assigned partner-level mentor—his development leader (DL) in McKinsey terminology—with whom he met quarterly to review his overall career progress and offer advice on his personal and professional development. In discussions with his DL Bray first raised the possibility of spending some time abroad. Recognizing this as an excellent way to build on his telecom knowledge spike and to broaden his general consulting skills, the DL made the arrangements for the young associate to relocate to the London office on a short-term loan.

It was when he became affiliated with the European telecom practice that Bray probably received his most intensive coaching. The practice leader, Michael Patsalos-Fox, took the Australian under his wing and encouraged him to work on Practice Development documents that would force him to tighten his thinking while simultaneously developing his influence. He then offered Bray the opportunity to present some of his ideas to senior client management, coaching him on how to communicate the key points most effectively. Patsalos-Fox also asked Bray to become engagement manager on several major studies he directed, helping his protégé develop his ability to focus and motivate people. And Patsalos-Fox opened the opportunity for the young consultant to act as adviser to engagement teams in other parts of Europe, helping build his network and broaden the roles he was able to play.

By the time Bray came up for election to partner, he had developed all the skills he needed to meet McKinsey's tough criteria. The in-depth knowledge he had acquired in the telecom industry made him an active contributor to practice development ("snowball making," as insiders referred to it), while the problem-solving, people-management, and networking skills that his

coaches and mentors had instilled gave him the client development (or "snowball throwing") ability that was vital at partner
level.

As a partner, Bray found he needed to work on developing a
very different set of skills and abilities. The watershed events of
the early 1970s had burned into the firm's collective conscious-
ness a recognition that too much attention paid to the client
development task (for which young McKinsey consultants were
so well prepared) could be very risky for the firm's long-term reputation. As Fred Gluck had emphasized, partners needed to
broaden their perspective and focus their primary attention on
coaching and developing the flood of young consultants being
recruited into the firm—more than five hundred a year by the
mid-1990s.

Warwick Bray seemed to understand very well the new challenges he faced as a partner. In 1996, when Patsalos-Fox felt it
was time for him to hand off his responsibility as head of the
European telecom group, he asked his young Australian protégé
to become coleader of the practice. Asked about his priorities,
Bray immediately responded that the first would be to attract,
develop, and retain the best associates.

Having experienced such intensive coaching and guidance during the previous eight years, it became quite natural for him to
assume this role of developing people and relationships within
the telecom practice. But even at the partner level, Bray continued to receive intensive feedback and support from senior directors to help him develop the mentoring skills so vital to his new
role.

Equally powerful was the way in which the firm's leadership
used its own example as a means of coaching newer partners in
their vital developmental role. Rajat Gupta, who succeeded Gluck
as the firm's managing director, spent more than half his time on
personnel issues, a good piece of which was coaching young partners. He used his high visibility to create a firmwide event called
the Practice Olympics. Teams of young associates from around

the world presented new concepts and ideas they had developed, with finalists facing off in front of a judging panel that included Gupta. Although he clearly could not build relationships with or offer counsel to all the firm's four thousand associates worldwide, his hope was that by acting as a role model in such high-profile forums, he could underscore the importance of the coaching role among the partnership.

As Gupta discovered, the skills required to lead McKinsey were different again from those that made a good entrepreneurial frontline consultant or even an effective partner-level coach. The role of organizational leader required an ability to balance the need for fit and alignment that allowed efficient ongoing operations with the opportunities and excitement created by maintaining a sense of dynamic disequilibrium within the organization. This delicate balance was difficult to achieve, since it required an exceptional ability to make excitement and challenge an integral part of the work environment while simultaneously inspiring the trust and belief in the institution that gave people the stability and confidence to stretch for the challenge.

This set of institution-building skills was also generally developed through the careful coaching and support of the few individuals who seemed most likely to succeed at them. Fred Gluck was one such individual. Almost from the moment he joined McKinsey from Bell Labs he showed that he was someone with visionary ambitions. Ron Daniel, managing director at the time, recognized this passion in Gluck and asked him to take firmwide responsibility for intellectual capital development.

With Daniel's encouragement and support, Gluck developed that role into one that essentially made him associate managing director for practice development. Through this process Gluck was able to build the personal skills and organizational relationships that later helped him bring about a major cultural change in this conservative firm. His ability to develop inspiring conceptual insights (the strategic imperative of developing knowledge, for example, or the need to view the firm primarily as a means of developing world-class people) then to offset this with hard-headed motivating challenges (his drive to calibrate client service by conducting postengagement audits to measure teams' impact) were largely skills developed working intensely with Daniel.

Gupta, too, revealed many of these institution-building skills when he took over as managing director, on one hand reinforcing the stability and commitments in the "one firm" philosophy, while at the same time legitimizing new thrusts into India, China, and Brazil and launching six special initiatives designed to drive knowledge development to a new level.

In the end, a great deal of McKinsey's enduring strength is due to the fact that the partners recognize that their business is "more of a craft than a science." Unlike many of their competitors who invest in developing tools and techniques and then in training their consultants in their use—the belief in the "science" of management—McKinsey partners have remained wary of packaged concepts.

Their belief in the craft of consulting has led to an extraordinary investment in personal coaching to develop the skills necessary to meet their standards for creative yet disciplined frontline consultants, development-driven principals, and institution-building directors. McKinsey's huge pool of candidates for entry positions and its thorough recruiting process ensure that almost all new consultants have the native intelligence, motivation, and personality to succeed in their roles. And their recruitment from elite institutions ensures that they have or can quickly obtain the knowledge and expertise they need. Yet only one in five will make it to partner, and fewer than half the partners will become directors. The statistics point to two things—McKinsey's extraordinarily high standards, and, despite the huge investment in coaching and mentoring (by one estimate, consuming 10 percent to 20 percent of the average partner's time), the difficulty of developing the key skills that are the best indicators of an individual's likely success.

Yet for both the individual who develops the key skills and the coach whose nurturing, demanding, and cajoling provides the catalytic energy, the achievement is a very satisfying one. As one senior partner described it, "The [mentoring] relationship begins as an act of will, but becomes much more of an emotional attachment over time. . . . You have to convey so much more than problem-solving skills and your personal network—you need to convey aspirations, instill values, build excitement and create a view that almost anything is possible."

One of his younger advisees responded, "If you know you have someone supporting you, you have much more confidence. You don't have to stay in your comfort zone—you can operate in your mentor's much larger zone." It is operating in that expanded comfort zone that allows individuals to develop vital new skills.

FAREWELL TO "ORGANIZATION MAN"

The reason the old Russian doll model of management worked well in the past was that it focused all managers on the core tasks that the classic divisionalized hierarchy was designed to achieve— to accumulate, distribute, measure, and control scarce capital resources. Reinforced by a behavioral context of compliance, control, contract, and constraint, this structural form and its supporting management model ensured the predictable and controllable behavior of "organization man."

As companies eventually discovered, however, the tight alignment of behavior to support top management's capital allocation decisions eventually led to strategic stagnation. Far from needing to subjugate individual differences by requiring conformity to a generic model of management, companies must recognize that in a knowledge-based environment, diversity of perspectives, experiences, and capabilities can become an important organizational asset.

At the heart of the emerging concept of the Individualized Corporation is a fundamentally different belief that companies can and must capitalize on the idiosyncrasies—and even the eccentricities—of people by recognizing, developing, and apply-ing their unique capabilities. It takes skill and sensitivity to see the potential in a struggling new recruit, and courage and patience to let him or her develop the unique capabilities he or she brings. But in the end, the personal and organizational rewards can be huge. Ask Red Auerbach.

Managing the Transformation Process: A Blueprint for Corporate Renewal

9

Since Earth's last encounter with Halley's comet, successive waves of downsizing, delayering, restructuring, and reengineering programs have swept through corporations around the world. Indeed, there must be few companies that have not tried to reinvent themselves in some form over the past decade—some more than once. Yet for every successful corporate transformation, there is an equally prominent failure. GE's dramatic performance improvement stands in stark contrast to the string of disappointments and crises that have plagued Westinghouse; ABB's ascendance to global leadership in electrotechnical businesses only emphasizes Hitachi's inability to reverse its declining fortunes; and Philips's turnaround since 1990 only highlights its own confused meanderings in the preceding ten years.

In the course of our research, we have studied more than twenty companies that have implemented innumerable programs intended to rationalize their inefficient operations, revitalize their ineffective strategies, and renew their tired organizations. We have sought to understand why some made progress in the difficult and painful battle for transformational change while others only replaced the dead weight of their bureaucracies with change program overload.

In observing how the successful corporate transformation processes have differed from those that struggled or failed outright, we were struck by two distinctions. First, successful transformation processes almost always followed a carefully phased approach that focused on developing particular organizational capabilities in an appropriate sequence. And second, actual transformation occurred only when the structural reconfiguration was reinforced by real and enduring change in the behaviors of individuals within the organization.

This chapter sets forth a general process model describing how companies have evolved from traditional divisionalized hierarchies to self-renewing Individualized Corporations by sequentially developing the individual capabilities described in part 2—entrepreneurship, learning, and self-renewal. And it will illustrate how this is achieved through the development and interaction of the company's behavioral context, its organizational processes, and its redefined individual competencies.

A PHASED SEQUENCE OF CHANGE

The problem with most companies that have failed in their transformation attempts is not that they tried to change too little but that they tried to change too much. Consider the events we observed over a nine-month period in one company. In the aftermath of a major restructuring, the new CEO embarked on a series of visioning retreats. One outcome was a senior management–endorsed definition of the company's core competencies that was then handed to a task force to recommend how they might be more effectively developed and managed. Meanwhile, the newly appointed chief knowledge officer launched an initiative to help the company become a more effective learning organization. And in a separate but contemporaneous initiative, consultants were called in to help design a reengineering program.

Sound familiar? Although this scenario describes the activities at one particular company, it could have been one of many. In their desperate search for more effective organizational models, managers have launched a flurry of unconnected programmatic activities in almost random order. Little wonder that in many companies frontline managers are completely bewildered in the face of the multiple and inconsistent priorities being imposed on them.

In contrast, the companies that were most successful in transforming themselves into more flexible and responsive organizations seemed to have much clearer understanding of what they were trying to achieve and pursued a sequence of actions that were extremely intense but comparatively simple. Although inevitably implemented in an intuitive manner and in a very company-specific way, when compared retrospectively, the transformational paths of these successful organizations looked remarkably similar.

Building on the simple but powerful ideas that Jack Welch used to describe the series of changes he implemented at General Electric, we developed a model that seemed to capture the transformational experience of several of the companies we were studying, including ABB, Motorola, Komatsu, AT&T, and Corning. Our underlying assumption is that the performance of any company depends on the strength of each of its component units, as well as on the effectiveness of their integration. (The model applies equally to the integration of individually strong functional groups along an organization's value chain, the synergistic linking of a company's portfolio of business units, or the global networking of its different national subsidiaries.) We used this simple yet fundamental premise to define the two axes of the corporate renewal model represented in Figure 9.1.

As they face the renewal challenge, most companies find themselves with a portfolio of operations that can be represented by the circles in Figure 9.1. They may have a few strong independent units and activities represented by the tightly defined but separate circles in quadrant 2. In addition, they usually have a cluster of better integrated operations that are not performing well individually, as depicted by the looser, overlapping circles in quadrant 3. And finally, most companies contain a large cluster of business units, country subsidiaries, or functional entities that don't perform well individually and are also ineffective in linking and leveraging each other's resources and capabilities. These units are illustrated by the ill-defined, unconnected circles in quadrant 1.

Most of the activity underlying recent transformation efforts has aimed at improving performance on one or both of the two dimensions represented. The overall objective of the process has been to move the entire portfolio of entities into quadrant 4, where individually strong units work together to create competitive advantage none of them could achieve independently.

Figure 9.1
The Phased Process of Corporate Renewal

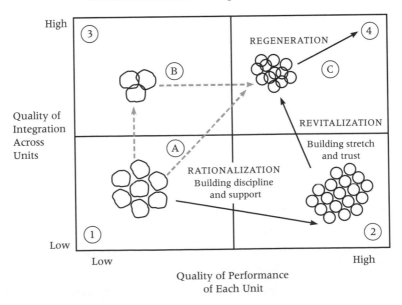

Some companies—General Motors during the 1980s, for instance—tried to improve performance on both dimensions simultaneously—an approach represented by diagonal path A. While intellectually and emotionally appealing, this bold approach typically produces internal contradictions and complexities that block effective change. GM discovered this during the 1980s when it pressured its five auto divisions to boost their individual market share and profitability while simultaneously improving cross-unit synergies. However, as management discovered the demands of coordinating body styling and chassis design often conflicted with the divisions' ability to differentiate and respond to their own individual market segments. Unsurprisingly, the effort failed.

Other companies' transformation efforts have pushed integration first, on the assumption that better synergy among units would help each one in its attempts to improve its individual performance. This change model, represented by path B, was taken by Philips in the mid-1980s. In a bold reorganization, company president Cor van der Klugt declared the company's consumer electronics, professional electronics, electronic components, and lighting businesses as "core interdependent" operations and tried to create structures and processes that would help them each manage their perceived interdependencies. However, it proved extremely difficult to integrate operations that were individually struggling with enormous internal difficulties. Even where they succeeded, the linkages connecting uncompetitive individual businesses primarily served to reinforce the liabilities of each. As corporate performance continued to decline in the late 1980s, the gallows humor among Philips managers held that "four drunk fat guys do not make an effective team."

The third transformational option, path C, is the one that follows the changes that have taken place at ABB over the past eight years. In the first phase, Barnevik stripped away much of the old bureaucratic superstructure to focus attention on the task of realigning the organization around the efficient operations of 1,200 frontline companies. It was at this stage that ABB energized hundreds of latent frontline entrepreneurs like Don Jans and gave them the mandate to build their businesses as if they owned them.

As the entrepreneurial engine of frontline units began to restart growth, Barnevik moved the company into the next stage of its transformation process, creating an integrated learning organization. In this phase, he focused the organization more intensely on the task of linking and leveraging the valuable resources and expertise that had been developed in pockets of entrepreneurial initiative throughout the company. Through such initiatives, ABB was able to combine products and technologies from dozens of its operating companies in Europe, Asia, and the United States and deliver the final turnkey power plant project in India or China.

By the mid-1990s, Barnevik was pleased with the progress that ABB had made, but knew the process was not complete. As it headed into the closing years of the century, ABB was entering into the self-regeneration phase of its transformation process. This was the stage in which the organization would have to learn how to balance the tensions and manage the paradoxes implicit in the new corporate model. It required managers to strive for superior individual unit performance while capturing the corporatewide benefits of cross-unit integration. Moreover, they had to do so in an organizational environment in which the hard-edge demands for operational efficiency were offset by the uplifting challenge of innovative expansion—a management model we described as "cooking sweet and sour." Like the first two stages, the revitalization phase would rest on the achievement of profound changes in the perceptions and behaviors of the people in ABB.

To explore in more detail how this multiphased model is managed through changes to a company's hardwired strategy structure and systems, as well as to the softer elements of culture, values, and norms, consider the evolution of the most celebrated corporate transformation of our time—the rebirth of General

Electric under Jack Welch's leadership. While most other highly diversified companies were breaking themselves up, Welch was proving that effective corporate management could add enormous value by improving the performance of individual businesses and achieving multibusiness synergies. In fact, his success in fundamentally redefining the way GE worked was widely recognized as the reason behind the company's 1,155 percent increase in market value in fifteen years.

GE is the appropriate company with which to illustrate the simple transformation model for several reasons. First, because it so clearly represented the old hierarchical corporate model, GE's commitment to becoming an Individualized Corporation is particularly powerful. Second, because Welch initiated these dramatic changes earlier than most, GE provides a better opportunity to observe the longer term impact of the transformational process. And third, because of the first two reasons, GE has become a widely accepted benchmark of best practice and Welch himself has become a model leader of such changes.

PHASE 1: RATIONALIZATION
EMBEDDING ENTREPRENEURIAL DRIVE

When he became CEO in 1981, Jack Welch inherited an organization that was widely regarded as one of the best-managed large industrial corporations in the world. This was the company whose creation of the strategic business unit (SBU) concept had provided a template for companies worldwide; its planning processes, long the standard in the industrial world, set the pace with the pioneering application of sophisticated strategic portfolio planning tools; and its development of group- and sector-level management to control the growing portfolio of SBUs also was widely imitated by large companies worldwide. In short, GE was the benchmark for large diversified corporations—the modern-day standard bearer for the professionally managed divisionalized hierarchy.

Furthermore, Welch took over the reins from Reg Jones, perhaps the most admired company leader in the country, a man *Fortune* magazine described as "a management legend." It would have been easy to stand on the legend's shoulders and refine the GE innovations that had become so widely accepted as leading-edge practice. But that was not Welch's style. Almost immediately, he began fretting that the company had too many under-performing businesses and too much organizational bureaucracy. Rather than building on GE's past accomplishments, Welch began restructuring the company's businesses, reconfiguring its organization, and redefining its management processes. Welch seemed to confirm *Fortune*'s prediction that the GE board had chosen "to replace a legend with a live wire."

In an analogy he was to use repeatedly, Welch saw the need to base the transformation of GE not only on changes to its "hardware"—its existing business strategy, organizational structure, and management systems—but also to its "software," which he saw as the values, motivations, and commitment of its employees. Although he employed both elements in the rationalization process, it was clear that his years of training in the GE system gave him a strong early bias toward changing the "hardware."

Changing the Hardware

Taking over in the midst of a recession and in a competitive environment in which Japanese companies were inflicting huge damage in many industrial sectors, Welch decided that his first priority was to focus attention on improving the operating performance and strategic competitiveness of GE's portfolio of businesses. His simple standard was to make each business number one or number two in its global market to achieve quality and performance that were "better than the best." To those who did not meet the standards, he said, "Fix it, sell it, or close it."

It was a bold move that radically reshaped the company's strategic portfolio. During the 1980s, GE sold or closed businesses that had represented about one-quarter of the company's 1980 sales. In the same period it was selling off assets worth almost $10 billion, the company made acquisitions worth $17 billion to reinforce the competitive position of the number one and number two businesses it retained.

As the business-level organization began to grind into action under his new challenging strategic criteria, Welch became increasingly frustrated with the layers of hierarchy and the density of corporate staff that insulated him from the business leaders who were emerging. Gradually he began to strip away the structures that his predecessors had so carefully built. By 1985, sectors, groups, and SBUs had been eliminated and GE's once powerful two-hundred-person corporate planning staff had been decimated. "As we develop our strategies," said Welch, "I want general managers talking to general managers, not planners talking to planners."

As Welch pushed to reduce the company's nine levels of management to his objective of four, the company's classic hierarchy, reflecting the old span-breaking theory that a manager should have no more than seven direct reports, was discarded. Believing the right number to be ten, fifteen, or perhaps even twenty, Welch deliberately created an organization that forced his senior-level executives to delegate more and more.

The changes were designed to redefine the management roles and relationships that had became embedded in the old hierarchical model. In particular, Welch wanted GE's operating-level managers to develop their roles around what he defined as ownership, stewardship, and entrepreneurship of the company's portfolio of competitive businesses. To drive ownership, he delegated more responsibility and pushed decision making deeper into the organization; to encourage stewardship, he decentralized more assets and resources and challenged his managers to leverage the company's return on them; and to spark entrepreneurship, he encouraged those on the front lines to initiate more action and to take more risks.

But the organizational impact of these strategic and structural changes was as traumatic as it was dramatic. Between 1981 and 1984, GE's total workforce was reduced by over seventy thousand. In this environment of restructuring, delayering, and downsizing, many began to feel overloaded and stressed out. And to Welch's chagrin, his aggressive restructuring earned him the nickname "Neutron Jack."

Changing the Software

Like most managers of his generation, Welch had a strong bias for the traditional levers of strategy, structure, and systems to drive change. But as powerful as they were, he found them to be blunt instruments. Gradually, his perspective began to soften and broaden. "A company can boost productivity by restructuring, removing bureaucracy, and downsizing," he said, "but it cannot sustain high productivity without cultural change."

By the mid-1980s, Welch's talk about number one and two businesses gave way to a passion "to combine the strength, resources, and reach of a big company with the sensitivity, leanness, and agility of a small company." It was an objective that demanded massive cultural change, but as Welch was to discover, changing "the smell of the place" is a much slower and more subtle task than changing a company's structure and strategy. Like most transformational leaders we studied, Welch soon found he could not change all the elements of his organization's embedded culture at once.

In the first-stage task of creating entrepreneurial drive, the bedrock element of behavior context was self-discipline, for it was this cultural norm that gave top management the confidence to empower those on the front lines. Even from his earliest days as CEO, Welch was building this cultural value at GE. Breaking with the GE tradition of framing strategies in detailed analyses and complex objectives, he set his operating-level managers the simple yet demanding challenge of making their businesses number one or number two in their global industry, then got out of the way.

By removing the layers of hierarchy that institutionalized the culture of compliance, Welch created an environment in which managers were required to set their own standards and evaluate their own performance. It was what he referred to as the mirror test: "Only you know whether or not the excellence is there," he told his operating managers. "Are you setting the standards of excellence? Are you demanding the very best of yourself?" Rather than just meeting the numbers, Welch wanted his managers to "face reality, see the world in the way it is, then decide for yourself how to act."

As he discovered, however, a single-minded focus on tough targets and self-imposed discipline can leave the organization stressed out and exhausted. After reaching that point in the mid-1980s, Welch became more conscious of the need to soften his tendency to push the organization to what he himself described as "the point where it almost comes unglued." He had to supplement his focus on discipline with an equal emphasis on support.

One key reason why frontline managers had became so overburdened and stressed out was that they had little or no experience in managing the resources or handling the responsibilities that had been pushed down to them. Recognizing this, Welch began to focus on ways in which he could support them in taking on their new tasks. He redirected the highly regarded Crotonville, New York, training facility from its traditional role of providing general management education and training to take on a more targeted agenda of designing and delivering company-specific organization development programs.

Equally powerful were the radical changes Welch made to GE's legendary strategic-planning and budget-development processes to become the basis for supportive discussion rather than forums for formal review. At a 1985 officers meeting, he asked each of his business managers to present their strategies in five one-page charts that defined the global market structure of their businesses, their key competitors' positions and expected actions, and their own business strategy and its expected impact. The subsequent give-and-take strategy discussions focused on how Welch and his key staff could help the business leaders achieve their objectives. It became an ongoing dialogue based on a five-page playbook and stood in sharp contrast to an annual exercise in reviewing and approving two-inch-thick plans in three-ring binders. Welch described his role in such discussions:

> It is the business leader's job to create and grow new businesses. Our job in the executive office is to facilitate. . . . Probably the most important thing we can promise our business leaders is fast action. When our business leaders call, they don't expect studies, they expect answers.

Gradually, Welch began to change management norms and practices, but always in a fairly controlled top-down manner that revealed his own history and experience as a manager who had cut his teeth on a classic hierarchy. It was only several years later that he really began to let go, and through a powerful process he called "Work-Out" he allowed frontline employees to redefine the behavioral context. In a series of carefully defined forums, those working under the constraining and controlling structures and systems were given the opportunity to challenge their managers with their problems and proposals for change. Through this process the norms of self-discipline and support were operationalized through hundreds of specific proposals and projects. After scores of such meetings throughout the company, empowerment had meaning, discipline and support became real, and "the smell of the place" began to change.

Changing the Behaviors

A subtle but important change was occurring during these first few years of realignment of GE's structural "hardware" and cultural "software." Consciously or unconsciously, Welch began shifting the management rhetoric and focus from a near obsession with structure and systems to impose direction and ensure compliance to a reliance on embedding changes in the attitudes and behaviors of individual organization members in order to build the desired change into day-to-day activities. And the behavior he was most trying to create within GE was a sense of entrepreneurial drive and initiative.

Heading up a company that was increasingly reliant on the motivations and self-initiated actions of hundreds and eventually thousands of people deep in the organization required Welch to focus much more intensely on human resource issues—employee development, management education, evaluation and reward systems, and so on. For example, twice a year he spent half a day with each of his business managers reviewing the human resource potential within their operations and quizzing the business leaders on how they were developing the high-potential people.

He also wanted to signal clearly the kind of entrepreneurial behavior he sought by reinforcing it through the reward system. While the company's predictable 3 percent to 4 percent salary increases supplemented by 10 percent to 15 percent bonuses to most at senior levels had clearly reinforced the old expectation of long-term loyalty, dependable compliance, and strong implementation skills, Welch wanted to recognize and support different behavior. He started acknowledging true corporate entrepreneurs with salary increases in the 10 percent to 15 percent range, bonuses of 30 percent to 40 percent to many fewer managers, and stock options that he began distributing to hundreds of effective frontline managers rather than continuing the practice of reserving them for the top echelons.

In the end, the new structural framework and behavioral context he was creating were designed to do nothing short of redefine the basic relationship between the company and its employees. Welch explained:

> Like many other large companies, GE had an implicit psychological contract based on perceived lifetime employment that produced a paternal, feudal, fuzzy kind of loyalty. That contract

has to change. . . . The new contract is that jobs at GE are the best in the world for people willing to compete. We have the best training and development resources and an environment committed to producing opportunities for personal and professional growth.

This organization of highly disciplined frontline competitors armed with the training and resources to allow them to take on the world was a very different model than the classic divisionalized hierarchy that had defined management behavior and relationships for decades. Like many companies in the first stage of the transformation process, GE was going through a period that many described as "inverting the pyramid." By empowering and energizing those running the company's portfolio of businesses and focusing them outward on customers and competitors, this change represented a powerful first step toward building the new organization model. (Figure 9.2.)

Figure 9.2
Inverting the Pyramid

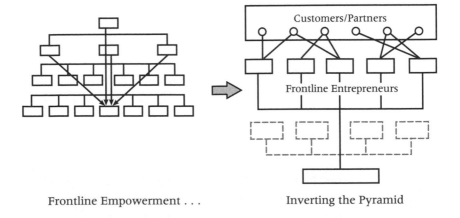

Frontline Empowerment . . .   Inverting the Pyramid

In the inverted pyramid, the fifteen viable frontline businesses became the company's basic structural unit, replacing the sectors and groups as the locus of power, and the behavioral context of embedded discipline and support supplanted imposed compliance and control, empowering those running the businesses to take charge. But as Welch and others discovered when they reached this stage, the transformation process was not yet complete. Some obstacles still blocked their way.

The first major problem derived from the fact that in the process of empowering those in the front lines, many in middle- and senior-level staff and line jobs (indicated by the dotted lines on Figure 9.2) were unsure of how to redefine their roles or reshape their relationships with the frontline entrepreneurs. In several transformation processes, these managers became "the layer of clay" that blocked further progress by trying to hold on to power. In some cases, they even subverted frontline initiatives to do so. For the transformation to succeed, the senior-level managers' new roles had to be defined.

The second major limitation of the inverted pyramid model was that each of the frontline entrepreneurs was focused on developing his or her own individual business opportunity. Even where customers, markets, or technologies overlapped, they had little incentive to cooperate. To remain viable, the company had to become more than a holding company for a portfolio of independent business, no matter how entrepreneurial they were individually.

PHASE II: REVITALIZATION
DEVELOPING INTEGRATIVE SYNERGIES

The primary focus of the first phase of GE's transformation process was on the restructuring, delayering, and downsizing activities that were designed to cut costs and enhance productivity. Little thought was given to growth or expansion. Indeed, GE's sales increased by only 8 percent during Welch's first five years at the helm, while its operating profits soared by 58 percent.

The risk, as some companies have discovered, is that the organization can become so "lean and mean" that it is unable to restart the engines of growth. Such problems arise when a company downsizes itself into a state of organizational anorexia that not only leaves it too weak to rebound but also too psychologically obsessed with cost-cutting to begin looking for new ways to expand.

Welch was careful not to allow GE to be sucked into that downward spiral. By the mid-1980s, as he grew increasingly frustrated with the nickname "Neutron Jack," he made it clear both inside and outside the company that he did not want GE to become known as a "lean and mean" organization. Instead, he wanted managers to strive to become "lean and agile."

The agility he sought was achievable in part by an organization in which resources and responsibility had been shifted down to more empowered frontline business managers. But, in Welch's view, it also required an organization in which resources, information, and expertise moved more rapidly and easily across internal and external structural boundaries. It was a concept he would later call "boundaryless behavior"—another profound shift he achieved through changes in the organization's "hardware" and "software."

Changing the Hardware

By the latter part of the 1980s, most of the radical changes to GE's structures and systems had been made. The new stripped-down framework was a great deal clearer and simpler, and operating-level managers were responding well to the openness and freedom it gave them. Yet, with the removal of the group and sector levels, the company had lost a major part of its ability to integrate across organizational units. Unless the company was to operate in a fragmented and compartmentalized manner, some kind of replacement coordinative mechanisms were required.

As he moved the primary focus from rationalization to revitalization, Welch began to create more cross-unit integrative forums to allow business-level managers the opportunity to share resources and transfer ideas. His first step was to create a Corporate Executive Council (CEC) consisting of his three-man office of the CEO and the managers with direct responsibility for the company's business. For two days each quarter, the CEC gave the top line managers the opportunity to grapple with common problems, share insights, and offer advice and assistance to each other on key problems and opportunities. For example, when the appliance business manager reported a major recall due to serious problems with compressors, his colleagues in the turbine and aircraft business were able to offer valuable technical support to solve the problem.

To push the opportunity for cross-unit integration deeper into the organization, Welch convened officers' meetings that pulled together the top hundred managers. Again, he encouraged them to find ways to work together more effectively. The relationships soon extended beyond the CEC and officers meetings, and individual managers began finding ways in which they could work together through internal sourcing (the plastics business increased its content in GE refrigerators to sixteen pounds a unit) or by working with common customers (as occurred in France when a joint venture in aircraft engines was able to help GE's medical business access the government-controlled health-care system).

Welch also wanted to redefine the role of the staff groups remaining at corporate headquarters "from checkers, inquisitors, and authority figures to facilitators, helpers, and supporters of operating line managers." He wanted each staff manager to ask, "How do I add value? How do I help people on the line become more effective, more competitive?" Again, his objective was to create an organization in which "ideas, initiatives, and decisions could move quickly, often at the speed of sound—voices—where once they were muffled and garbled by being forced to run the gauntlet of staff reviews and approvals."

But while he was successful in creating a more open and supportive learning environment at the top levels of GE, it was only at the end of the decade that Welch succeeded in pushing the process deep into the organization. One of the most powerful tools was an integrative process called Best Practices. Begun as a staff-led analysis of high-productivity companies, the program's site visits to selected benchmark organizations were soon recognized as offering extremely powerful learning opportunities. To spread that learning through the organization, the company created a Best Practices development program at Crotonville. Every month, a dozen people from each of GE's businesses would exchange views and experiences as they compared their own management approaches with the best practices they studied. It was in this stimulating and collaborative environment that cross-unit learning took firm root.

One situation Welch liked to recount to inspire imitation involved GE's Canadian appliance company, which had successfully adapted the flexible manufacturing approach of a small New Zealand appliance maker. Welch told how two hundred managers and workers from the giant Louisville plant toured the Canadian operation to learn what they had done. After a series of Work-Out sessions, the U.S.-based team put plans in place to reduce the production cycle time of its appliances by up to 90 percent while simultaneously increasing product availability. The achievements at Louisville were so impressive that it immediately became a popular internal Best Practices destination for teams from other businesses, from locomotives to power-generating equipment. It was classic Welch—another inspirational story designed to spark widespread imitation.

Changing the Software

As important as the various informal structural overlays were in creating new integrative channels and forums, Welch recognized that unless he could change GE's slow-moving bureaucratic culture, the cross-unit processes he had framed would never become effective. Over time, he continued to shift his attention from the "hardware" elements that had so preoccupied him in the early and mid-1980s to focus most of his time on the "software" issues of GE's culture, values, and management style.

One of the strongest signals of this new emphasis came early in 1989, when Welch began to talk about the need for the company to develop a management approach based on speed, simplicity, and self-confidence. These themes provided him with the means to expand on the behavioral context that he had begun to build earlier by supplementing the norms of discipline and support with characteristics described in chapter 6 as trust and stretch.

In the aftermath of downsizing and delayering, one of the key challenges for Welch personally and for the company in general was to rebuild a trust relationship with GE's somewhat traumatized employees. Welch knew that people would be willing to take risks only if they trusted top management. And only if they trusted each other would they be able to work together. Welch's new management values spoke directly to those needs. "Becoming faster is tied to becoming simpler," said Welch. "And on an individual and interpersonal level, this takes the form of plain-speaking, directness and honesty."

Through his widespread interactions with organization members, Welch personified this open, candid management style that he believed was the only way he could overcome the negativism and mistrust attached to his "Neutron Jack" reputation. He met tirelessly with GE employees, from his regular one-on-one reviews with business leaders to his no-holds-barred exchanges with over five thousand managers a year at Crotonville sessions. He participated in Work-Out sessions, traveled tirelessly to visit GE facilities worldwide, and, above all, found every opportunity possible to communicate his message. In the process, he gradually earned not only the respect of his employees but also their trust.

The process took years. When Work-Out was first introduced, for example, many were suspicious that this was just another exercise designed to justify cuts and layoffs. Welch immediately and unequivocally assured the organization that this was not so. By carefully controlling Work-Out sessions so they were not abused, Welch was able to use the process to create a new level of openness and understanding between employees and their bosses—albeit through the often uncomfortable process of openly confronting reality for the first time. Again, trust was being built.

In an environment characterized by fairness, where people learned to trust their bosses and their colleagues sufficiently to take risks, the other key element of a growth-supportive context is stretch. Welch's notion of self-confidence addressed this need directly as he encouraged his managers to free themselves from the confines of their box on the organization chart, share information freely, listen to those around them, then move boldly. "Shun the incremental," he told them. "Go for the leap."

As the company moved into the 1990s, this notion became an increasingly important one for Welch, to the point where he believed that self-imposed stretch targets could drive the company's growth much more effectively than management-imposed budget numbers. He began to preach a belief that "budgets enervate, stretch energizes." He explained:

> We used to timidly nudge the peanut along, setting goals to increase operating margin from 8.53 percent to 8.92 percent, for example. Then we'd engage in time-consuming, bureaucratic negotiations to move the number a few hundredths one way or the other. The point is, it didn't matter. . . . Today, we challenge the organization with bold stretch targets, and trust that everyone will do as well as they can to achieve them. People come in with numbers far beyond what we would have asked for. . . . It would kill accounting professors but it's real.

Welch believed the power of this approach allowed the company to break out of its long history of single-digit operating margins as it stretched for the 15 percent target set in 1991. In 1995,

GE reported an operating margin of 14.4 percent, a little short of the objective, but well beyond anything it could have achieved under the lowest-common-denominator negotiations that characterized the old budgeting process.

Changing Individual Behavior

Once again, Welch took his grand strategic objective of reigniting growth and his lofty organizational goal of creating more cross-unit learning and translated them into the changes they implied for individual GE employees. Building on the focus and drive that he had instilled in his managers in the earlier rationalization process, in the revitalization phase he shifted his attention to the need for self-confidence and collaboration.

"Boundaryless behavior" was the antiparochial, anti-incremental style required if GE was to link and leverage the pockets of entrepreneurial activity it had created so effectively in the early and mid-1980s. After years of work to change the internal norms, there was more than a hint of pride in Welch's 1966 declaration of "the demise of the 'Not-Invented-Here' syndrome within GE." By the mid-1990s he was proudly recounting a long list of innovations and ideas that had been transferred across its businesses, driving growth and profitability throughout the company.

Like ABB, Komatsu, Corning, and other companies at this stage of their transformation process, by the 1990s GE had begun to develop a set of entirely new management roles and relationships. The organizational configuration that shaped them was also evolving with the inverted pyramid of the mid-1980s, starting to look more like the integrated network of entrepreneurial activities described in chapter 4 (Figure 9.3).

If the first stage of the transformation process was focused on the task of releasing entrepreneurial hostages on the front lines of the organization, the second phase transforms middle- and senior-level managers into developmental coaches. Their key task is to stretch the individual entrepreneurs to become the best they

Figure 9.3

Building an Integrated Network

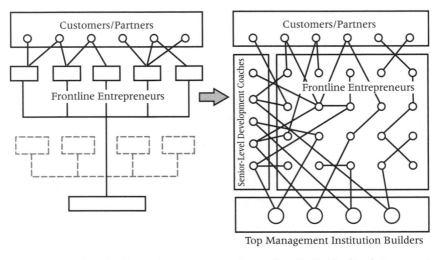

From Inverting the Pyramid . . .   . . . Toward an Individualized Corporation

can be and to create an open, trusting, and collaborative environment that facilitates boundaryless behavior.

Welch showed he was serious about this shift in the corporate culture at GE when he began talking about the consequence of being a "Type IV" manager—one who got results but did so without sharing the values of openness and collaboration. Calling this "the ultimate test of our commitment to the transformation of this company," Welch began to remove these individuals, even though they met their financial projections.

As his senior managers began supporting the ideas and initiatives of those in the front lines and the company developed an openness that encouraged collaborative behavior, Welch began to feel that the company was approaching the organizational objective he had set for it. GE was developing the soul of a small company within the body of a big one.

PHASE III REGENERATION:
ACHIEVING CONTINUOUS SELF-RENEWAL

Well into the 1990s Welch seemed to recognize that what he had been working on for over a decade was more than just the major shake-up and realignment of a grand old company. It was the creation of a fundamentally different corporate model operating through a radically different management philosophy. And the more this new model was defined and developed, the clearer it became how delicate and vulnerable it was during its slow and painful birth process.

Compared to the clarity, stability, and certainty of roles and relationships in the old divisionalized hierarchic model, the operating environment being created in the new Individualized Corporation seems much riskier. Do those on the front lines have the skills and motivation to use the resources and responsibilities thrust upon them in an effective manner? Can senior-level executives step back from the "take-charge" behaviors that allowed them to achieve their current positions and take on a more supportive coaching role?

The greatest risk occurs when organizations in transition face some kind of major challenge or crisis. Most employees are not only exhausted from the protracted and profound change process, they feel tentative and uncertain about their newly defined management roles. In such a stressful time, the natural inclination is to retreat into the comfort of the more predictable and stable old management model. Even with a well-designed multiyear change process, it is hard to change behaviors learned, practiced, and reinforced over a twenty- or thirty-year career.

The third stage of the transformation process must respond to that postchange letdown. The challenge is to create an internal environment that supports both individual initiative *and* team-based behaviors on a continuous basis. To achieve this objective, the new behavioral context and management roles must be reinforced long enough for the organization to free itself from the embedded practices and conventional wisdoms of the past. Thus the final objective is to build the company as a self-renewing organization, something that Welch described in his vision for the new millennium: "Our goal is to build a GE that renews itself constantly, exhilarates itself with speed, and freshens itself through constant learning."

As Welch would readily acknowledge, however, GE has not yet achieved this final stage of self-generated, continuous renewal. However, he does recognize occasional glimpses of such behavior. Many of GE's traditional industrial businesses, for example, have injected a new lease on life into old operations by shifting their focus to selling services to supplement their traditional product-focused approach. GE Medical Systems has begun offering service contracts on hospital chains' medical equipment, including competitive products; Aircraft Engines is reducing the cyclicality of its sales by signing engine maintenance agreements with major airlines; and Power Generation is pursuing opportunities to operate and maintain power plants. Yet management understands that in order for such promising regenerative initiatives to become the rule rather than the exception, the company will need to continue to change its organizational software and hardware.

Changing the Hardware

The further companies like GE evolve through their transformation processes, the more the hardwired, static structures recede to become little more than a general framework defining the overall configuration of the company's resources and responsibilities. Much more important in shaping the finer grained management decision making and action is the portfolio of processes that define the core workings of today's multidimensional and flexible organizations.

Elements of the new renewal process have already been put in place in the Best Practices programs that began decoupling managers from the comfort of their old ways of doing things, and even pushing them beyond the limits of internal cross-unit learning. By encouraging managers to seek out leading-edge practice wherever it may exist, Welch institutionalized a dynamic disequilibrium that is replacing the constant quest for fit and alignment with a model focused on embedding continuous challenge and stretch.

Although the shape of all the new process changes is not yet clear, some clues are appearing in the new challenges Welch is framing for the organization. He is identifying the need to raise quality standards yet another notch, and has pointed to Motorola as an inspirational standard; he has focused on the need for faster new-product introduction and has suggested that GE is some distance behind pacesetting companies like Hewlett-Packard and Toshiba; and he has thrown down the gauntlet in areas like globalization, information technology, and service, which he defines as "the biggest growth opportunities in our history."

Yet, unlike many of his earlier highly specified programs and challenges, these latest key goals seem remarkably broad and underdefined, perhaps in an effort to create space for the individual business leaders and their teams to take the leadership on this next round. All the signs seem to point to the fact that Welch is creating the context for the process of self-renewal to take root.

Changing the Software

Having created a management model based on speed, simplicity, and self-confidence, Welch confirmed its importance by aggressively rewarding those who personified those values and removing those who were unable to manage this way. As the organization moves into the third stage of its transformation process, the challenge is to retain the dynamic tension that exists in a behavioral context characterized by stretch, discipline, trust, and support. Like most companies, GE had developed a system that had a strong bias toward hard-edged objectives and measures, and managers found it a great deal easier to develop and sustain the norms of stretch and discipline than to embed the values of trust and support.

Welch understood better than most the need to offset the constant hard-edge demands and pressures it took to create a truly competitive company with the softer-edged values that maintained energy and motivation among employees. While he set high standards and stretching goals for his organization, he also made sure there was constant celebration of the milestone achievements—even when the organization didn't quite make it, as in the case of 1995 operating margin of 14.4 percent. More fundamentally, while he removed the barriers and guarantees that had long protected GE employees from the competitive reality of the outside world, he also offered them the comforting assurance that the company would give them the training and support to become the best they could be. By maintaining this yin-yang tension, Welch was able to redefine GE's relationship with its employees in a way that gave them the responsibility for ensuring the company's continued success.

Like Dr. Maruta at Kao, Welch was learning that creating a self-renewing organization required management to cook sweet and sour—to offset the hard-edged context of discipline and support with the softer elements of stretch and trust. Only by building rationalization and revitalization into the ongoing flow of business operations could a company prevent the kind of accumulation of inefficiencies that call for massive restructuring programs. And only through constant renewal can they stem the gradual atrophying of initiative and collaboration that stalls out growth and requires jump-starting through expensive revitalization initiatives.

Changing the Behaviors

To develop a truly self-renewing organization, companies must do more than change the exhibited behaviors of key management groups. They must ensure that such changes are rooted in the personal values and beliefs of all members of the organization. It was this notion that ABB's chief operating officer, Goran Lindahl, described as his core task—to turn engineers into capable managers, and managers into effective leaders. "When we have developed all our managers into leaders," he explained, "we will finally have a self-driven, self-renewing organization."

That leadership model of behavior is most rapidly communicated if it is modeled effectively by those at the very top of the organization. Their ability to spend more time framing context and less time defining content becomes a critical example for middle-level managers who must learn to manage more through coaching and supporting than by directing and controlling. And their ability to see themselves not just as the chief strategists but also as the institution builders reinforces to those on the front lines that they do not just work for a company, they also belong to an organization.

FROM CATERPILLAR TO BUTTERFLY

Managers of most large companies around the world have recognized the need to make some radical, even transformational changes to their well-embedded organizational and management models. Yet few have gone as far as GE in throwing off their familiar old ways, and most have done little more than tinker at the margins.

The problem is that their mind-sets are so dominated by structural and engineering models built into their traditional heirarchies that they try to bring about change by reconfiguring the assets and reengineering the processes rather than focusing on how to change individual motivations and interpersonal relationships. Those who have successfully led their companies through transformational change have all discovered one central truth: No change will occur until people change.

But as the GE case illustrates (and the other examples described throughout this book confirm), the voyage from a structural hierarchy to a self-renewing, Individualized Corporation is likely to be long and painful. The metaphor of a caterpillar transforming itself into a butterfly may be romantic, but the experience is an intensely unpleasant one for the caterpillar. In the process, it goes blind, its legs fall off, and its body is torn apart to allow beautiful wings to emerge. We cannot deny the pain involved, but for companies that succeed, the wings they grow as they develop the new behavioral context will allow them to take flight as Individualized Corporations.

Part 4

TOWARD A NEW
CORPORATE ERA

A New Moral Contract: Companies as Value-Creating Institutions

10

There is much truth in the saying that every living practitioner is a prisoner of the ideas of a dead theorist. Distrustful of theory and bonded to the "real world" as they are, corporate managers have nonetheless become unwitting victims of a set of a ideas that have run out of explanatory power.

Much of modern management is based on research and theories of corporate behavior that were rooted in the era that ranged from the trust-busting days of the early decades of the century to the post-Vietnam period of profound pessimism about people and institutions in general. Often framed by economists who disliked and distrusted companies, these theories have collectively created an amoral philosophy of management premised on a highly instrumental relationship between the company and the society on the one hand, and between the company and its employees on the other. In their day-to-day choices and actions, most managers work within the framework of these theories, more out of unconscious conformity to established norms rather than any conscious understanding of or agreement with their underlying assumptions or logic.

Underneath the structures and processes of the Individualized Corporation lies a very different management philosophy. Grounded in a different set of assumptions about human nature and the roles of institutions in modern society, this philosophy leads to some very different beliefs about the role of the company in society, about the relationship between employers and employees, and about the functions of management and its obligations as a profession. Overall, it posits a very different moral contract between the individual, the company, and society. It is this new moral contract, more than any of its specific operating characteristics, that is the essence of the Individualized Corporation.

Creating Value for Society

It was the economists' distrust of the motivations and actions of corporations that led to the nationalization movement in Europe and the regulatory environment in the United States. On both sides of the Atlantic, economists supported these broad movements with well-developed theories of how companies distorted the beauty of open markets and pure competition by erecting barriers and obstructing the free flow of resources. In this imaginary war, battle lines were drawn between companies on one side and markets on the other.

It did not take long before industrial organization economists and the business strategists who followed them saw the opportunity to turn these findings on their head. If social welfare was served by preventing companies from impeding and obstructing competition, then it stood to reason that companies could enhance their position by maintaining or, even better, raising, obstacles to competition.

Companies as Value Appropriators

Thus in Michael Porter's highly influential strategic theory, a company finds itself in the midst of a set of competing forces that pit it not only against its direct competitors but also against its suppliers, customers, and those who may become its future competitors. Management's core challenge, he maintains, is to tighten the company's hold over its suppliers and customers and to find ways to keep existing and future competitors at bay, protecting the firm's strategic advantages and allowing it to benefit maximally from them.

The essence of this theory is simple: The objective of a company is to capture as much as possible of the value that is embodied in its products and services. The problem is that there are others—customers, suppliers, and competitors among them—who want to do the same. As the economists point out, if there is genuine, free competition, companies can make no profits above the market value of their resources. The purpose of strategy, therefore, is to prevent such open and free competition: to claim the largest share of the pie while preventing others from eating your lunch, to mix metaphors.

Implicit in the economists' model is the assumption that by preventing open and free competition, the company impedes social welfare. Yet this view simply does not stack up against the reality of modern societies. The last hundred years have seen an uninterrupted and unprecedented improvement in the quality of human life, due, in large measure, to the ability of companies to continuously improve their own productivity and their talent for creating new products and services. As Nobel laureate Herbert Simon noted, to call modern society a "market economy" is a misnomer; it is primarily an "organizational economy" in which most economic value is created not through the economists' ideal of atomistic competition in a completely free market but within efficient, well-functioning corporations involving large numbers of people acting collectively, coordinated by the broader purpose of the total organization.

The truth is, contrary to the dominant theory, most companies do not usurp markets to appropriate value for themselves at the cost of social welfare. Rather, in healthy economies, successful and prosperous corporations coexist with intensely competitive markets in a state of vigorous and creative tension with one another. Companies create new value for society by continuously creating innovative new products and services and by finding better ways to make and offer existing ones; competitive markets, on the other hand, relentlessly force the same companies, over time, to surrender most of this value to others. In this symbiotic coexistence, companies and markets *jointly* drive the process of creative destruction that Joseph Schumpeter, the Austrian economist, showed to be the engine that powers economic progress in capitalist societies.

The problem with Porter's concept of companies, which has shaped the thinking of a generation of managers, is that it is based on a static view of the world, in which the size of the economic pie is given. In this zero-sum world, all that is then left to be decided is how the pie is to be divided up, and corporate profits must indeed come at a cost to society. Schumpeter's very different view of companies is based instead on a dynamic analysis of how the pie gets bigger in a positive-sum game in which there is more for all to share. In this view, instead of merely appropriating value, companies serve as society's main engine of discovery and progress by continuously creating new value out of the existing endowment of resources.

Companies as Value Creators

The contrast between these two views of a company comes sharply into focus in a comparison of the management approaches of Norton and 3M, or Westinghouse and ABB. Managers at Norton and Westinghouse lived in the zero-sum, dog-eat-dog world of traditional strategic theory. When they found the market for a product too competitive for them to dictate terms to their buyers and suppliers, they sold those businesses. And when they found a company that had created an attractive new product or a business, they bought it. Their primary management focus was on value appropriation—not only vis-à-vis their customers and suppliers but also vis-à-vis their own employees. Recall Robert Kirby's claim when he was Westinghouse's CEO that he would fire his own mother if she wasn't doing her job.

At 3M or ABB, a very different management philosophy was at work. While Norton tried to develop increasingly sophisticated strategic resource allocation models, 3M's entire strategy was based on the value-creating logic of continuous innovation of new products and new technologies. And the same business that Westinghouse abandoned as mature (read: not enough opportunity for value appropriation), ABB could rejuvenate in part by investing in improving productivity and by incorporating new technologies.

As these companies created new products or new markets, society rewarded them with high margins. However, over time the margins eroded, as competitors caught up, and what the companies lost in profits their customers specifically and society more generally received in the form of additional value. By the time most of the initially high profit margin had been squeezed out of any given innovation through market pressures, the companies had discovered new opportunities in the form of new products and applications.

The difference between these companies is not just that 3M and ABB focused on innovation and improvement while Norton and Westinghouse did not but that they based their actions on some very different beliefs about what defines a company. At Norton and Westinghouse, managers thought of their companies in market terms: They bought and sold businesses, created internal markets whenever they could, and dealt with their people with market rules. Through the power of sharp, marketlike incentives, they got what they wanted. People began to behave as they would in a market—with an acute sense of self-interest.

What happens, however, when each individual acts only in his or her self-interest is that a company loses its essence as an institution of modern society—the essence of what distinguishes it from a market and, thereby, endows it with the ability to create new value in a way that markets cannot. In a market, people carry out economic exchanges only when each of the parties involved can clearly see how he or she, individually, gains from such a transaction. Because markets have no purpose or vision of their own, they can ruthlessly weed out inefficiencies by reallocating resources among the best available options. But, for the same reason, markets are not very good at creating innovations that require new combinations of resources.

By thinking of their companies in market terms, Norton and Westinghouse became the victims of market logic. All they could do was strive for squeezing more efficiencies out of everything they did. Their strategy focused entirely on productivity improvement and cost-cutting. They were unable to innovate not because they were physically incapable of doing so but because the logic of the market they adopted internally did not allow for creation beyond the efficiency of existing activities.

To create innovations and new value, a company must typically allow a level of slack—that is, it must sacrifice some efficiencies—by allocating resources to uses that do not yield the highest immediate returns. This is because even path-breaking innovations often begin at a disadvantage to existing alternatives and reach their potential only over time. By thinking of their companies in market terms, Westinghouse and Norton became victims of the market straitjacket, unable to pursue the dynamic efficiencies of innovation and value creation because of their total focus on achieving static efficiencies of current productivity.

Visions like ABB's purpose "to make economic growth and improved living standards a reality for all nations throughout the world"; values such as Kao's espoused belief that "we are, first of all, an educational institution"; and norms like 3M's acceptance that "products belong to divisions but technologies belong to the company" all emphasize the nonmarketlike nature of the Individualized Corporation, encouraging people to work collectively toward shared goals and values rather than restrictively, within their narrow self-interests. They could share resources, including knowledge, without knowing how, precisely, each of them would benefit personally—as long as they believed that the company overall would benefit. Ultimately, this philosophy distinguishes these organizations from companies that think of themselves as markets and allows them to create innovations through a spirit of collaboration.

Companies like 3M, Kao, and ABB offer their employees a temporary respite from market forces by actually muting the market's sharp incentives and creating an atmosphere more supportive of collaboration and sharing. In so doing, they create a (temporarily) protected environment in which individuals can work together to challenge market forces and generate new combinations of resources, therefore creating new value for society. This is precisely what 3M does by allowing people 15 percent "bootleg" time to pursue their own projects. A company's ability to constantly create new value for society is a product of a management philosophy that espouses viewing the company not just as an economic entity but also as a social institution that allows individuals to behave more cooperatively and less selfishly than they would in the economist's free market.

Companies and Society

Over the twentieth century, corporations have earned an enormous amount of social legitimacy, which has been both a cause and a consequence of their collective success. Amid a general decline in the authority of other institutions—political parties, churches, the community, even the family unit—corporations have emerged as perhaps the most influential institution of modern society, not only in creating and distributing a large part of its wealth but also providing a social context for most of its people, thereby acting as a source of individual satisfaction and social succor.

Yet, in the closing decades of the century, corporations and their managers suffer from a profound social ambivalence. The evidence is everywhere—in Bill Clinton's White House conference on corporate responsibility in response to public outcry at the wave of downsizing in the United States, in Tony Blair's reviews of the role of the corporation in the United Kingdom, in the deep suspicion of large companies in France, Korea, and even Germany, and in the public furor over executive pay in every country where the astronomical wealth of entertainers, entrepreneurs, sports people, or even independent professionals raises few eyebrows. In fact, in most countries, corporate managers have been knocked from the pedestal on which they once sat to emerge as one of the least trusted constituents of society.

In spite of a few visible misdeeds, this perception may be objectively unfair, weighed against the huge material benefits companies have given to society. Yet, the perception persists—and it is potentially one of the greatest risks that corporations face today. The clear lesson from history is that institutions decline when they lose their social legitimacy. This is what happened to the monarchy, to organized religion, and to the state. This is what will happen to companies unless managers accord the same priority to the collective task of rebuilding the credibility and legitimacy of their institutions as they do to the individual task of enhancing their company's economic performance.

Far from thinking of their companies as agents for destroying social welfare, most managers we have met believe that their primary role is to create value. Their guilt lies in their unwillingness to explicitly confront the question of what role their companies play in society or to consciously articulate a moral philosophy for their own professions. Through this act of omission, they have left others—economists, political scientists, sociologists—to define the normative order that shapes public perceptions of themselves and their institutions. Those perceptions, in turn, have seduced many managers into thinking about their companies in overly narrow terms, in the process becoming unconscious victims of the value-appropriation logic and losing their ability to create new value for society.

This is why individuals like Percy Barnevik and Yoshiro Maruta will earn their places in history—not because of their influence on the economic performance of their firms, because hundreds of managers achieve that routinely, but because they have championed a corporate philosophy that explicitly supports the view of companies as value-creating institutions of society. And, they have reshaped the organization and management processes of their companies around this new philosophy to give birth to a new corporate form, the Individualized Corporation.

In doing so, they have created a new moral contract between the company and its constituencies that not only is more satisfying for managers but is also more effective in protecting and growing their companies. The problem of a strategy of value appropriation is that ultimately it is self-defeating. It is like a strategy of holding back the tide; like the tide, the ability of others to overcome a company's defenses cannot be held back forever. With such a strategy, the company is squeezed ever more tightly into a corner, with each round of value appropriation consuming ever more effort, until finally there is no value left to appropriate. Companies that think of themselves as a market ultimately succumb to the market—as happened in Norton's acquisition by St. Gobain and by Westinghouse's dismemberment under CEO Michael Jordan. Hansen Trust followed a classic value appropriation strategy, as did ITT under Harold Geneen. Ultimately, each of these companies fell victim to the same market logic they had so enthusiastically embraced within themselves. In that process, they actually destroyed value for their customers, shareholders, and employees. In contrast, 3M and Kao continue to grow profitably, spawning new products and businesses, creating customer satisfaction, employee enthusiasm, and shareholder wealth, and ABB continues to expand and strengthen its leadership position in its businesses, at times by acquiring the spent-up parts of companies like Westinghouse and rejuvenating them with the power of its very different philosophy.

CREATING VALUE FOR PEOPLE

Within the concept of a company as a value-appropriating economic entity, its relationship with its own employees is also shaped by appropriate norms. Like all other constituencies, people become a source from which the company can extract value to achieve its economic objectives. At its worst, this appropriation philosophy leads to a ruthless exploitation of workers. But at least in countries with an infrastructure of employment laws and some form of external labor markets, it more often translates into something usually regarded as benign—a relationship based on employment security. In this well-established, implicit contract, the company guarantees the employees jobs in exchange for their willingness to execute the tasks allocated to them and to abide by the strategies and policies that management establishes for the company.

The Traditional Employment Contract

It is counterintuitive to think of the offer of employment security as exploitive. This is not how the arrangement emerged nor how, even now, most employees and employers think about it. Yet, as benign as it may appear, it is this relationship that has historically enabled companies to extract the greatest possible value out of their employees.

Unlike machines, people cannot be owned. Yet, like machines, they become most valuable to a company when they become specialized to the company's businesses and activities. The more specific the employee's knowledge and skills are to a company's unique set of customers, technologies, equipment, and so on, the more productive they and the company become. Without employment security, employees would hesitate to invest their time and energy to acquire specialized knowledge and skills that may be very useful to the company but may have limited value outside of it. Similarly, without any assurance of a long-term association, companies lack the incentive to commit resources to help employees develop such company-specific expertise. Employment security provides a viable basis for both to make such investments.

While the company benefits from such specialization directly, in terms of efficiency and productivity, it also benefits indirectly. The more company-specific an employee's skills, the less useful they are to anyone else. Not only does this make employees less mobile, it also reduces their market value. The result is that the company can pay less *and* demand loyalty.

Exploitative or not, this contract defined a viable relationship. Employees developed the special knowledge and expertise the employer's business needed, thereby enhancing the company's efficiency but also limiting their own skills and mobility. While companies absorbed the risks by granting them cradle-to-grave jobs, employees promised loyalty and obedience, which allowed companies to ensure that their strategies were implemented efficiently. For the same reason that so many companies adopted the model of the "Organization Man," they also adopted this psychological contract—to make their people as reliable and controllable as the other assets the company owned.

During the past decade, this implicit psychological contract has broken down. Company after company—not only in the United States and Europe but also in Brazil, India, Japan, and Korea—has pursued efficiencies through downsizing and outsourcing strategies that effectively ended the possibility of secure employment. From a stopgap measure to stem the flow of red ink to the bottom line, such strategies have since become standard procedure, used routinely by even the healthiest of companies. The ever-present threat that survivors of one outsourcing program will emerge only to be caught in the undertow of the next wave of head-count cuts has effectively made the traditional contract both nonviable and noncredible for both employers and employees.

Much of the blame for the breakdown of the old employment contract has been placed at the doorstep of greedy management. In truth, in some companies the process of downsizing has gone too far, has been carried out inhumanely, and has been motivated by extreme short-term considerations. Yet, in the end, it is not management but the market that has made the traditional contract nonviable. In a stable world, the old contract could work. Competitive advantage, once developed, could be sustained for long periods of time, as companies like IBM, Caterpillar, Kodak, and Xerox proved. In this world, top-level managers could determine the company's strategy, specify what they wanted employees to do, and define the skills they needed to do it. For their part, employees could gradually develop those skills through training and apprenticeships and use them productively in the service of the corporate strategy—often over their entire careers.

However, in a dynamic world, a source of competitive advantage in one period becomes not only irrelevant but also often a source of competitive disadvantage in another. Core competencies become core rigidities. Valuable knowledge and skills become rapidly outdated, often at a rate faster than many people's learning capacities. Markets shift, technologies change, prices erode, and new competitors render obsolete not only profitable products but whole business systems at a stroke. This is the kind of environment that most companies face today. In such conditions, the old contract is not just nonviable; any effort to pretend otherwise is immoral.

Consider the case of ABB. Historically, North America and Europe accounted for a vast majority of new power plant demand. Today the situation has changed, and by the late 1990s, China's requirements will greatly exceed those of the United States or the whole of Europe. While the company's resources are all in the North and the West—yesterday's markets—its opportunities are all in the South and the East. To face up to the future, the company has had to reduce employment in North America and western Europe by fifty-four thousand people, while building up a forty-six-thousand-person organization in the Asia-Pacific region, almost from scratch. This degree of geographic mismatch between resources and opportunities is a reality for many companies. In this light, and as markets and technologies change almost daily, to guarantee employment is to commit competitive suicide.

Unfortunately, as many major corporations have come to recognize, the alternative of a free-market hire-and-fire regime is not a viable replacement for the traditional employment contract. The same forces of global competition and turbulent change that make employment guarantees unfeasible also enhance the need for trust and teamwork that cannot thrive in a sterile environment of reciprocal opportunism and continuous spot contracting. Similarly, after the fifth job change in as many years, even the star bond trader often comes to recognize that the workplace is not just an arena for economic exchange but also a source of social engagement. While they may maximize their economic returns by continuously hiring themselves out as commercial mercenaries, most people also yearn for the sense of fulfillment that comes from belonging to an organizational family.

The New Moral Contract

To resolve this tension, the concept of the Individualized Corporation is grounded in a very different moral contract with people. In this new contract, each employee takes responsibility for his or her "best-in-class" performance and undertakes to engage in the continuous process of learning that is necessary to support such performance amid constant change. In exchange, the company undertakes to ensure not the dependence of employment security but the freedom of each individual's employability. It does so by providing employees with the opportunity for continuous skill updating so as to protect and enhance their job flexibility within the company and their opportunities outside. At the same time, companies operating under this new contract must strive to create an exciting and invigorating work climate, not only to enable the employees to use their skills to benefit the company's competitive performance but, more important, to retain them in the company even though they have the option to leave. Jack Welch articulated this contract thus when he described the new employment relationship at GE:

> The new psychological contract . . . is that jobs at GE are the best in the world for people who are willing to compete. We have the best training and development resources, and an environment committed to providing opportunities for personal and professional growth.

What we call the new moral contract (Figure 10.1) is more than a fresh spin on a company's old human resource policies to legitimize layoffs; it embodies a fundamental change in management philosophy. No longer are people seen as a corporate asset from which to appropriate value. Under the new contract, they are a responsibility and a resource to which to add value. Its adoption implies a rejection of the paternalism, even arrogance, that underlies lifelong employment contracts. It recognizes that only the market can guarantee employment and that market performance flows not from the omnipotent wisdom of top management but from the initiative, creativity, and skills of all employees. At the same time, however, it acknowledges that companies have a moral responsibility for the long-term security and well-being of the people they employ, and for helping them to become, in a phrase we have used before, the best they can be in what they choose to do.

This new moral contract also demands much from employees. It requires that they have the courage and confidence to abandon the stability of lifetime employment and embrace the invigorating force of continuous learning and personal development.

Figure 10.1

The New Moral Contract:
A Role-Responsibility Reversal

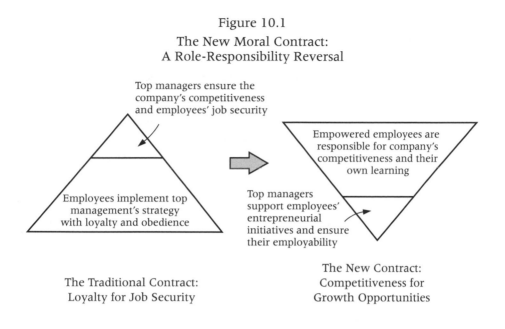

Top managers ensure the
company's competitiveness
and employees' job security

Empowered employees are
responsible for company's
competitiveness and their
own learning

Employees implement top
management's strategy
with loyalty and obedience

Top managers
support employees'
entrepreneurial
initiatives and ensure
their employability

The Traditional Contract:
Loyalty for Job Security

The New Contract:
Competitiveness for
Growth Opportunities

They must accept that the security that comes from perfor-mance in the market is ultimately both more durable and more satisfying than the security offered by a paternalistic manage-ment.

However, responsibility for performance is accompanied by uncomfortable demands. Those on the front lines are no longer able to wait for top management to legitimize unpleasant but necessary decisions. If assets can be reduced, employees closest to the operations must do it; if expenses are out of line, it is their responsibility to cut them; and if the work can be done with fewer people, the decision to increase productivity or reduce head count is also theirs. Rationalization can no longer be a once-in-ten-years cleanup job, driven by a top-down, corporatewide program. It must become an ongoing activity, a part of the con-tinuous "sweet and sour" cycle, managed by frontline employees. No one has emphasized the hard edge of the new contract more sharply than Andy Grove:

No matter where you work, you are not an employee. You are
in a business with one employer—yourself—in competition
with millions of similar businesses worldwide. . . . Nobody
owes you a career—you own it as a sole proprietor. And the
key to survival is to learn to add more value every day.

But, at the same time, companies must counterbalance this
hard-edged expectation with a reciprocal commitment to add
value to people. The need for significant investments in training
and development is only a part of it. These investments must be
made to protect and enhance the employability of individuals as
much as to increase the productivity and efficiency of the company; to support their broader general education and not just to
enhance their job-specific skills. As Anita Roddick of The Body
Shop says, "You can train dogs; we educate people."

Few companies take their commitment to the employability of
people more seriously than Motorola. In a context of radical
decentralization of resources and decisions to the divisional level,
employee education is one activity that is managed by Motorola
at the corporate level, through the large and well-funded
Motorola University that has branches all over the world. Every
employee, including the chief executive, must undertake a minimum of forty hours of formal coursework each year. Courses
span a wide range of topics, from state-of-the-art coverage of
new technologies to broad general management issues, allowing
Motorola employees around the world to update knowledge and
skills in their chosen areas. It is this commitment to adding value
to people that allowed Motorola to launch and implement its
much imitated "Six Sigma" total quality initiative. At the same
time, the reputation of Motorola University has increasingly
become a key source of the company's competitive advantage in
recruiting and retaining the best graduates from leading schools
in every country in which it operates.

More recently, Motorola has further upped the ante on its
commitment to employability by launching the "Individual

Dignity Entitlement" (or IDE) program. The program requires all supervisors every quarter to discuss six questions with those they manage. A negative response from any employee to any one of these questions is treated as a quality failure, to be redressed in accordance to the principles of total quality management. Yet, even at Motorola, which has invested more in its people than most companies and has long been a champion of employability, some units reported failures of more than 70 percent in the first round of IDE reviews. Beginning in 1995, the company began systematically addressing the negatives by identifying and then eliminating their root causes. This is the hard edge of the new moral contract on the management side—the commitment to help people become the best they can be that counterbalances the tough new demands on people that the "employability for competitiveness" contract creates.

Motorola's "Individual Dignity Entitlement" Program

1. Do you have a substantive, meaningful job that contributes to the success of Motorola?

2. Do you know the on-the-job behaviors and have the knowledge base to be successful?

3. Has the training been identified and been made available to continuously upgrade your skills?

4. Do you have a personal career plan, and is it exciting, achievable and being acted upon?

5. Do you receive candid, positive or negative, feedback at least every 30 days which is helpful in improving or achieving your personal career plan?

6. Is there appropriate sensitivity to your personal circum-
    stances, gender and/or cultural heritage so that such
    issues do not detract from your success?

What the New Contract Is Not

While we have described at some length what a moral contract based on employability is, it is important at this point to emphasize what it is not. First, it is not a catchy new slogan to free management from a sense of responsibility to protect the jobs of their people. At Intel, Andy Grove could make the kind of demands he did on people only because his own past actions had established beyond doubt the lengths to which he was willing to go to protect the interests of his employees. As we have seen, during the "memory products bloodbath" in the early 1980s, when every other semiconductor company in the United States immediately laid off large numbers of people, Grove tried all other expedients—the 10 percent rule, the 25 percent rule, the equity sell-off—to stem the flow of red ink before he finally resorted to lay-offs and plant closures. It is this kind of proven commitment to people that makes a contract based on employability credible and its hard-edged demands on people acceptable.

Another thing a moral contract is not: an altruistic agreement to educate and develop people at company cost so that they can find better jobs elsewhere. Paradoxically, the new relationship actually enhances a company's chances of retaining its best people. In a contract based on employment security, most people who lost their mobility because of overspecialization or skill obsolescence stayed with the company simply because they had no alternative. But, the very best people often left—frustrated by the constraints and controls that were the other side of the contractual relationship. In contrast, the promise of employability itself is a great motivator for people to remain with the company that makes it. They know that even if they can cash in their current employability at a premium, they run the risk of falling victim to the next round of skill obsolescence in a company that does not have the same commitment to adding value to people. The same broad and advanced skills that make people employable outside the company also make them more adaptable to different jobs and needs within it, thereby making it easier for the company to use their expertise more flexibly and in higher value jobs.

Finally, the contract based on employability is not some program that can just be installed. We cannot emphasize enough that it is the embodiment of a wholly different philosophy—one that requires management to create the excitement and satisfaction that bonds people to the company in a much healthier and more honest relationship than the dependency of guaranteed employment security. A moral contract based on employability goes hand-in-hand with a management commitment to empowerment that can be sensed in the new "smell of the place." The combination, then, leads to a durable and mutually satisfying relationship between the individual and the organization that the traditional employment contract abandoned. But by building the new company-employee relationship on a platform of value adding and continuous choice rather than on a self-degrading acceptance of one-way dependence, the new contract is not just functional: It is also moral.

BUILDING A "SHARED DESTINY" RELATIONSHIP

Is this notion of the modern corporation focused on creating value externally and internally what the British call "cloud-cuckoo land"? Is it all wishful thinking of wet-behind-the-ears softies who do not know how hard and unforgiving the world of business really is; or worse, of ivory tower academics who preach what they cannot do?

This book is filled with examples of highly successful companies that practice the philosophy of value creation. Kao, 3M, Intel, and Canon are all textbook examples of companies that earn healthy profits year after year by continuously focusing on the task of creating value for society rather than expropriating value from it. Canon made its own highly successful laser printer technology obsolete by inventing and then aggressively promoting the bubblejet on the grounds that its functionality-to-cost ratio yielded higher value to customers; Intel fueled the information revolution by relentlessly following "Moore's law," creating the next generation of chips that allowed its customers to do new things, while at the same time wiping out its own earlier generation of products; and Kao decided to enter the cosmetics industry and use its advanced technology to create the high-functionality Sofina range to compete with overpriced mediocre products in expensive jars. In each of these companies, value creation was both the stated objective as well as the proven outcome.

Without a moral contract based on employability, McKinsey and Andersen Consulting could not be in business. Recruiting the very best talent is the number one key success factor in the consulting industry. Yet these firms can give partnerships to only one in five of all they hire. The rest must leave the company. The promise of employability—and proven ability to deliver on that promise—is the only reason that fresh graduates all over the world, well knowing the high job risks, still seek to join them.

We began our description of the Individualized Corporation with the story of Don Jans's rediscovery of management as a counter to those who believe that companies cannot teach old dogs new tricks. So we conclude our discussion of the new managerial philosophy committed to value creation with another story—the remarkable history of Unipart, a struggling U.K.-based parts manufacturer, and the way in which it transformed itself under this powerful management philosophy.

The Unipart Story

Unipart was born out of the residue that resulted from the dismemberment of the chronically sick, government-owned British Leyland, the U.K.-based automobile company. The company's most valuable assets and resources soon found other homes. Jaguar was initially floated as a public company and then bought by Ford; DAF of the Netherlands acquired BL's trucks division; the bus business went to Volvo; and British Aerospace purchased the car division, renaming it Rover. What was left was a rump of a totally uncompetitive parts business—an inefficient manufacturing operation and a portfolio of units engaged in sourcing and distribution. It was this unpromising residual business that became the Unipart Group of Companies following an investor and employee buyout in 1987.

At its founding as an independent company, Unipart suffered a two-to-one handicap in productivity vis-à-vis Japanese auto parts companies and an astonishing hundred-to-one gap in quality, according to a U.K. Department of Trade and Industries study. It inherited an extremely confrontational work climate, the product of a heavily unionized workforce crossed with a traditional and autocratic management. Furthermore, the company's adversarial relationships extended to its suppliers and its customers—principal among them being Rover, earlier a sister unit within British Leyland.

By 1996, the story had changed utterly. Turnover had doubled to nearly £1 billion, ($1.6 billion) profits had quadrupled to over £32 million ($50 million), and a recent Department of Trade and Industry study had announced that Unipart was the only U.K.-based company in its business to meet world-class standards on quality.

Behind the company's transformation was one man, Unipart CEO John Neill, and his absolute commitment to what he called "a shared destiny relationship." The philosophy was not an ex-post rationalization or a politically correct way to celebrate the success, but was stated clearly and firmly by Neill in 1987 as the fundamental principle on which the company would function. At that time he said:

> We have made a mess of our industry. The short-term power-based relationships have failed us. Many Western companies still believe that it is a superior way to secure competitive advantage. I think they are absolutely wrong. . . . We must create shared destiny relationships with all of our stakeholders; customers, employees, suppliers, governments and the communities in which we operate. It is not altruism, it is commercial self-interest.

Neill's words were more than empty rhetoric. His objective of creating "shared destiny relationships" was soon converted from philosophical ideal into implementable action, in all the company's external and internal relationships. A classic example of how Unipart fundamentally changed its external relationships is provided by its dramatic shift away from the traditional purchasing-agent-driven, price-oriented contract negotiations that previously dominated its supplier relationships. Under the "shared destiny" philosophy, Unipart management determined that it had to deepen and strengthen those linkages, adding value to the partnership by helping the supplier succeed.

A good example was Unipart's relationship with Tungstone, a hundred-year-old U.K. battery producer. Once enjoying a 25 percent share of the U.K. replacement battery market, by 1989 Tungstone had declined to the point where the company seriously considered exiting the business. Just one symptom of its deeply rooted problems: In 1989, fewer than 50 percent of its shipments to Unipart, one of its largest customers, met the stipulated lead time of two to three days.

It was at this point that John Neill convinced John Richardson, Tungstone's CEO, that his company should join Unipart's "Ten(d)-

to-Zero" program. Instead of functioning as a traditional supplier rating system, Ten(d)-to-Zero emphasized the measurement of joint Unipart-supplier performance across ten criteria from transaction costs and lead time to defect rates and delivery errors. A joint Unipart-Tungstone employee team was assigned responsibility for each of the performance dimensions, creating a joint learning process to drive down the measures to a zero score. The objective was to use the shared expertise and improved feedback to create quality improvements and cost reductions that both Unipart and the supplier would share.

Through innovations in such areas as forecasting, electronic data interchange, and production processes, the "Ten(d)-to-Zero" partnership began generating significant and rapid performance improvements in cost, quality, and on-time delivery. Norman Finn, Tungstone's director of sales and marketing and the principal link in the relationship with Unipart, described the process this way:

> We've seen other companies who talk about strategic alliances, but Unipart is the only one of these companies that really takes these things seriously. Shared destiny relationships are not easy, and require an incredible amount of hard work every day, and at times, we were exchanging information which even I was uncomfortable sharing with a customer. It is a process of continuous attention to the very smallest of details and a joint commitment to continuous improvement. We did not stop after improving on-time product availability from 48 percent to 80 percent in eighteen months, for example, and today we are generating results in the range of 96–97 percent.

And Unipart found that its shared destiny philosophy had a multiplier effect, adding value beyond its own constituent relationships. For example, Tungstone's Finn, whose attitudes had been shaped in the traditional adversarial buyer-seller relationships of the U.K. automobile industry, became an enthusiastic disciple of the new approach: "We began taking many of the things we learned in the relationship with Unipart to our other customers," he said. "Although not everyone was ready for it, we have worked hard to break through the trust barrier. . . . Everyone had to change; I had to change. But now I'm at the point where I believe there are widespread benefits to be gained from what Unipart calls a shared destiny philosophy.

Within Unipart, the shared destiny philosophy was having an equally powerful impact. It was redefining the relationship between the company and its employees in a way that would add value by helping them achieve their potential rather than being pigeonholed and stuck in place as they had so often been in the past. One of the most visible commitments to the new approach came with the creation of Unipart University. Built at a cost of £2 million ($3.5 million), and led by Dan Jones, coauthor of the best-selling book *The Machine That Changed the World,* the university was a key element of John Neill's strategy to make the company world-class. Neill personally taught regular courses there and described it as a place where "learning was a priority, where theory met practice, where colleagues became teachers and collaborators, and where fear of change evolved into confidence and capability." He saw its role as the spiritual center of Unipart, where people could absorb the company values and develop and diffuse a common language.

Unipart University was also the place that visibly embodied the company's commitment to a shared destiny relationship: Its doors were open to all of Unipart's constituencies and its courses were routinely attended not only by company employees but also by people from customer and supplier organizations, government officials, directors and employees of companies interested in learning from Unipart's approach, and members of the local community.

Any employee of the company could walk into the university's information technology center, sit in front of a computer, and learn a little or a lot, with help from a trained and friendly staff. Laptop computers were available on loan from the center, and a wide range of software from traditional spreadsheets to personal finances and even gardening programs was available for checkout from the university library. Regular courses covered topics such as continuous process improvement and providing outstanding customer service, as well as special sessions on topics ranging from team building to problem solving. Any employee could self-nominate to most of the courses.

One employee who frequently nominated herself for courses was Judith Harris, a personal assistant whose career development into a product manager's job illustrates the new energy and commitment Unipart was able to release in its once sullen workforce. Almost from her first day in the company, she sensed the opportunity to develop her potential. "There was a certain vibrancy about the place," she said. "There was always someone you could call on, or a course you could take . . . and, as long as you were up front about things, you were encouraged to try new things. It was okay to make mistakes."

Recognizing her potential, Harris's first boss encouraged her to join a four-year program in management studies and gave her a half day off each week to attend classes. The rest of the staff in the group covered for her if emergencies required attention when she was away. Then, halfway through her course, the company supported her interest in discovering other career options by allowing her to spend a few days working in other departments for a firsthand look at alternatives. Through this process, Harris identified a position in the marketing services team to which she aspired. On the basis of a strong recommendation from her boss, she landed the job even though such a role was rarely filled by someone with her limited background and experience. As she described the move,

> Suddenly I had left this safe, secure position and I was thrown into the fast-paced area of product promotions. I had

three months to develop a brand-new product promotion. I found out quickly what was meant by a team effort. Everyone worked together to help launch the spring promotion.

In 1996, when an opening for brake products manager came up, Harris had the motivation, the knowledge base, and the skills needed for the job. Recalling her evolution from a personal assistant to a middle-level managerial position, she described the incredible personal and professional changes she had experienced since joining Unipart: "I've actually gone beyond anything I ever dreamed of doing before. I used to be too frightened to express my own views. That has changed. Now, I'll try to do anything; I have the confidence. And it's not just at work—its spilled over into my personal life."

John Neill's vision for Unipart was not born of the economist's narrowly defined model for achieving profitability through the zero-sum game of extracting value from others. Rather, it was based on a much more expansive positive-sum value-creation perspective that we found much more typical of successful managers in the companies we studied. Like Neill, these executives had enormous faith in what they could create by engaging, energizing, and empowering their constituencies to work together for their mutual benefit. The result of the collective actions of these executives is something rather more than they thought: the emergence of a more realistic and assertive philosophy of the role of the corporation as both a powerful economic entity *and* an important social institution using its economic power to add value to society generally and to people's lives individually. Clarifying and confirming this new moral contract between the Individualized Corporation and its constituencies remains one of the greatest challenges for today's corporate leaders. The credibility of their positions and the legitimacy of their institutions depends on their ability to do it.

The Changing Role of Top Management: Beyond Strategy, Structure, and Systems to Purpose, Process, and People

11

In the years following the last visit of Halley's comet, as scores of the world's most prominent companies started to stumble and fail at about the same time, a rising Greek chorus of doom began to interpret their widely publicized problems as a sign that the era of the large corporation was coming to a close. The evidence seemed compelling. In the United States, corporate icons like IBM, Sears, and General Motors were in dire straits; once mighty European companies like Olivetti, Volkswagen, and Philips were even more embattled; and even seemingly invincible Japanese giants such as Hitachi, Matsushita, and Mazda began to falter badly. Such behemoths, the argument went, were simply too big, too slow, and too inflexible to adapt to the new business environment. They were corporate dinosaurs whose time had passed and were assuredly headed for extinction.

These arguments seemed persuasive, particularly as huge companies like ITT, ICI, and Westinghouse began to break themselves up voluntarily, while numerous others suffered the same fate at the hands of Wall Street asset strippers. Even more intriguing was the buzz about a radical new corporate model comprising a fluid network of small agile companies working as a so-called virtual corporation.

Yet, in the midst of these prophesies of doom, we found that many large global companies were adapting successfully to the emerging business environment without undergoing drastic strategic dismemberment or radical organizational reincarnation.

Beyond reconfiguration of assets and resources, the realloca-tion of roles and responsibilities and realignment of goals and objectives are much more profound yet subtle changes that frame the behaviors of these companies. It is this difference in organiza-tional and managerial practices that drives the organization to become an Individualized Corporation. Encouraged in part by a response to changing environmental and competitive needs but also, in part, by the convictions of a set of top-level managers who have learned that the old managerial doctrine that they mastered in the course of their own successful careers is no longer sufficient, companies are learning to respond to the demands required by their new role in society.

In order to implement the new corporate philosophy, what in the roles and tasks of top management must change? How must their own personal orientations and their explicit or implicit beliefs be transformed along with these role changes? And, even more important, who must the top management become so as to embody and embed the new corporate philosophy within their own organizations?

BEYOND STRATEGY, STRUCTURE, SYSTEMS

In examining the histories of the companies we were studying, it became clear that the more closely linked they were to the last big corporate revolution, the more likely they were to have adopted the management philosophy born of that change. The creative genius of Alfred Sloan and Pierre Du Pont was to create a structural framework that not only facilitated their diversification strategies but also established a radical new management practice. Based on the delegation of responsibility to a new level of general managers at the division level, this new practice was made possible only by the development of sophisticated information and planning systems that allowed corporate staffs to keep control of their diverse operations. Reinforced by the success of these companies and the flood of imitators, this basic approach eventually became the core doctrine of modern management. Structure follows strategy, it preached. And systems support structure. This strategy-structure-systems management doctrine led to the development of a highly effective corporate model, which was ideally suited to an era in which the most critical task was for top-level managers to allocate the company's limited financial resources to the competing opportunities proposed by the various division general managers, then monitor performance against their plans by means of a sophisticated set of control systems.

But as corporations race toward the twenty-first century, a fundamental shift is undermining the foundations of the strategy-structure-systems doctrine. Knowledge is replacing capital as the critical scarce resource, and management is being challenged to create an organizational environment that can develop, leverage, and diffuse this valuable new competitive asset. At the most basic level, by trying to understand this change in terms of the old strategy-structure-systems paradigm, many companies have found that they were able to deal only with the symptoms of their problems, not their root causes. More fundamentally, in order to manage the transformational change implied, managers have had to recognize the need for an entirely different management philosophy. Adding to the difficulty was the fact that such recognition requires them to change their own core roles and responsibilities in some basic and profound ways.

The companies that were successful have done just that. They have adopted an entirely different management philosophy. They place less emphasis on following a carefully analyzed strategic plan than on building an energizing corporate purpose. They have developed their organization less through changes in formal structure and more through developing effective management processes. And they have become less focused on managing the systems to control employees' collective behavior than on working directly to develop their capabilities and broaden their perspectives as individuals. In short, they have moved beyond the old doctrine of strategy, structure, and systems to embrace a broader, more organic philosophy built on the development of purpose, process, and people.

It is important to note that this new philosophy does not imply a complete rejection of the old doctrine; that is why the argument is deliberately framed as *beyond,* not *from,* strategy, structure, systems. Our observation is that in the emerging knowledge-based, service-intensive environment in which companies compete today, management must supplement traditional hard-edged tools with the perspective offered by the broader and more dynamic model that we have defined as the purpose, process, people philosophy. With this fundamental shift in philosophy management lays the foundation stone for building the Individualized Corporation.

Although this change in philosophy pervades the entire organization, its starting point is top management. Before it can bring about a corporatewide realignment in behaviors and beliefs, top management must first change its own priorities and mentalities. It must broaden its role as the designer of corporate strategy to one defined more as the shaper of a broad institutional purpose. The top level must expand its focus from architect of formal structure to builder of organizational processes. And it must expand its key task beyond managing systems to developing and molding people.

Beyond Strategy to Purpose

Since at least the time of Alfred Sloan, the powerful, even heroic, image of the CEO as omniscient strategist has been ingrained in business history and elaborated through management folklore. Few of today's practitioners have escaped hearing how Sloan conceived of the classic five-tiered product-market segmentation strategy that powered General Motors for over half a century. Today, this legendary image is reflected in and reinforced by the autobiographies of modern-day corporate leaders from Lee Iacocca to "Chain Saw" Al Dunlap.

When companies were smaller and less diversified and their operating environment was simpler and more predictable, the ability of those at the top to set a clear strategic agenda was undisputed. But as companies grew larger and their operating environment became more complex, senior executives needed elaborate systems and specialized staff to ensure that headquarters could review, influence, and approve the plans and proposals of their divisions and business units. Over time, the workings of the increasingly formalized planning processes eclipsed the utility of the plans they produced.

Ironically, disaffection only increased as senior managers ceded responsibility for unit-level strategy to division and SBU managers, and shifted their own attention to crafting an overall corporate-level strategic framework and logic. That shift led those at the top to explore a succession of new strategic concepts that would allow them to retain, or in many cases recapture, their historic role as the company's strategic leader. They experimented with the elusive concept of creating synergies among related businesses; they employed elaborate models to help define and balance cross-funded strategic portfolios; and in recent years they began to develop notions of a broad strategic vision or highly focused strategic intent as a means of providing an umbrella for their portfolio of businesses. But while each of these approaches yielded some useful insights, none provided top managers with the powerful strategic tool they were searching for to reestablish their control over strategy.

Meanwhile, those in the divisions and SBUs actually running the businesses grew increasingly confused about their roles. The elaborate contortions required to fit their strategies into a synergy-

304 / Sumantra Ghoshal and Christopher Bartlett

driven corporate rationale frustrated them; classification of their complex businesses into simplistic portfolio-funding roles demotivated them; and aligning around a vague-sounding strategic vision or an overly constraining definition of strategic intent left them cynical. All in all, rather than contributing to an integration of corporatewide activity, top management's attempt to provide strategic leadership often had the opposite effect.

The problem is rarely one of the CEO's personal inadequacy but rather the assumption that the CEO should be the corporation's chief strategist, assuming full control of setting the company's objectives and determining its priorities. In an operating environment where the fast-changing knowledge and expertise required to make such decisions are usually found on the front lines of each business, this assumption is untenable. Strategic information cannot be relayed to the top without becoming severely diluted, distorted, and delayed. And even when it does survive the journey in some useful form, top-level executives rarely have the current knowledge, the specialized expertise, or the fine-grained insight needed to make the sophisticated judgments implied by the strategic proposals.

Andy Grove, CEO of Intel, acknowledges that for a long time neither he nor the other top managers at his company were willing or able to see how the competitive environment had undermined Intel's strategy of being a major player in both memory chips and microprocessors. Yet for two full years before top management recognized that the company had to get out of the memory business, various project leaders, marketing managers, and plant supervisors were busy refocusing Intel's strategy by shifting resources from memories to microprocessors. As a result, Grove came to the somewhat humbling conclusion that neither he nor any of his top team had the control over the strategy they thought they had. He recalled the period:

> We were fooled by our strategic rhetoric. But those on the front lines could see that we had to retreat from memory

chips. . . . People formulate strategy with their fingertips. Our most significant strategic decision was made not in response to some clear-sighted corporate vision but by the marketing and investment decisions of frontline managers who really knew what was going on.

As Grove reflected on the lessons of the most important strategic decision in Intel's history, he acknowledged the need for top management to adjust its thinking about how it influenced corporate direction. Instead of concentrating their attention on refining the tools to help them develop strategy more precisely at the top, corporate leaders need to spend more time framing the environment that allows more strategic initiative and debate to emerge from below. Reflecting on future strategic shifts as dramatic as the exit from memories, he said:

The more successful we are as a microprocessor company, the more difficult it will be for us to become something else. . . . We need to soften the strategic focus at the top so we can generate new possibilities from within the organization.

Yet, while top-level managers are acknowledging their own limits as strategic gurus, many have learned that the people who can "formulate strategy with their fingertips" are deeply disaffected. Neither valueless, quantitative planning objectives nor mechanical, leveraged incentive systems nurture the sense of commitment or motivation necessary to foster initiative from within. And whatever fragile relationship exists is being eroded as successive waves of restructuring, delayering, and retrenching undermine any reserve of corporate loyalty.

In an environment in which people no longer know—or even care—what or *why* their companies are, leaders can no longer afford to focus only on refining the analytic logic that frames the strategic processes. Strategies can engender strong, enduring emotional attachments only when they are embedded in a broader organizational purpose. Today, the corporate leader's greatest challenge is to create a sense of meaning within the company, which its members can identify, in which they share a feeling of pride, and to which they are willing to commit themselves. In short, top management must convert the contractual employee of an economic entity into a committed member of a purposeful organization.

Managers can create an exciting and energizing sense of purpose in a variety of different ways. Recall Poul Andreassen, who transformed ISS from a small Danish office-cleaning company to a global leader in its business—not by articulating an ambitious growth target or through some energizing call to arms to leap ahead of a key competitor but by developing a sense of purpose: in particular, by creating an opportunity for his people to feel a sense of pride in their work. In an occupation that was historically seen as demeaning, he created personal dignity by asserting that daily office cleaning was a *professional* service, by backing up that assertion through positioning ISS as the benchmark for all service businesses ("like Xerox is to photocopying"), and by investing significantly in the technological basis of the business, in the quality standards, and most important, in the skills of his employees.

Similarly, Yoshiro Maruta created a sense of purpose at Kao by emphasizing the company's obligation to use technology to develop products that would be useful to society by improving the quality of life for ordinary people. The company's purpose also implied an even more radical notion—that Kao was primarily an educational institution, and that the most basic responsibility of every member of the organization was to teach and learn.

But of all the companies included in our research sample, none took this purpose-shaping role further than IKEA, the Swedish furniture company that since 1973 has been exporting its revolutionary concept of furniture retailing around the world. Reacting to the cartel-like agreements among Swedish furniture manufacturers and retailers that kept furniture prices beyond the reach of young people trying to set up their first homes, Ingvar Kamprad, the founder of IKEA, started his company in 1953 not only to exploit a business opportunity but also to tackle a social problem. As he said to us,

> A disproportionately large part of all resources is used to satisfy a small part of the population. . . . Too many new and beautifully designed products can be afforded only by a small group of better-off people. IKEA's aim is to change this situation. We shall offer a wide range of home furnishing items of good design and function at prices so low that the majority of people can afford to buy them.

Throughout the history of the company, Kamprad has emphasized a vision larger than IKEA. As he repeats in all his interactions within the company and outside, "Once and for all, we have decided to side with the many. We know that in the future we may make a valuable contribution to the democratization process at home and abroad." In turn, this purpose—of creating "a better everyday life for the majority of people"—has shaped the strategy of IKEA—an extreme focus on cost-consciousness, clarity about both product range and design parameters, and the location and layout of retail outlets, for example. It has also shaped the values of the company—embodied in the Swedish word *ödmjukhet*, implying humility, modesty, and respect for one's fellow human beings—and guided the selection and development of the people who have built IKEA's business.

Although the specific focus around which the sense of purpose was created differed widely in our sample of companies, there was one common element. In each case, the purpose transcended the company's narrow self-interest and was linked to broader human aspirations worthy of long-term pursuit. This is where the new societal role of companies described in the last chapter intersects with the changing role of top management that is the focus of this chapter.

The concept of strategy, both as it has evolved in academic thinking and as it has come to be adopted in corporate practice, reflects and reinforces the notion of companies as profit-maximizing agents of economic exchange in a large and complex environment. This minimalist, passive, and self-serving definition grossly underestimates the institutional role of corporations in modern society. As key repositories of resources and knowledge, corporations have become one of society's principal agents of change with a unique responsibility for creating and distributing wealth by continuously improving their productivity and innovativeness. Equally important is the role they play as the most important forums for social interactions and personal fulfillment for many in society.

It is this fundamental philosophical distinction that differentiates those top executives who see themselves simply as designers of corporate strategy from those who define their task more broadly as the shapers of institutional purpose. Purpose is the embodiment of a company's recognition that its relationship with stakeholders is defined in terms of interdependence rather than simple one-way dependence. In short, purpose reflects the moral response to a broadly defined sense of responsibility rather than an amoral response to a tightly focused sense of opportunity.

Beyond Structure to Process

In the post–World War II decades, a generation of managers grew up seeing firsthand how the divisionalized structure created by Sloan and Du Pont both facilitated the management of their increasingly diverse portfolios and institutionalized the strategy of diversification by supporting expansion into new businesses and markets. Not surprisingly, these managers became convinced that their control over the company's structural configuration was perhaps the most powerful tool they had to shape the broad direction of the organization and the drive of its members. Almost universally, structure joined strategy in the elite of tools so powerful they were reserved for top management's exclusive use and control.

By the early 1980s, however, it became apparent that while many of these structural innovations had indeed produced the desired expansion and diversification, the growth had come at some cost. The weight of increasing bureaucracy in the highly developed multidivisional model was crushing any sense of entrepreneurial initiative; its fragmentation and compartmentalization of resources and capabilities was inhibiting knowledge diffusion and organizational learning; and the hierarchical relationships so perfectly designed to ensure the efficiency and provide the controls necessary to refine existing operations proved incapable of creating and building new businesses internally.

As they began to reshape their greatly simplified organizations, many top-level executives began to focus their attention on building integrative processes that would contribute more directly to the company's strategic value added. In its most drastic and radical form, the shift from a structure-based to a process-based organization model manifested itself in the reengineering programs that swept through many companies in the first half of the 1990s. While many of these massive exercises created value as they obliterated old structures and rebuilt work flows around scores of outcome-oriented processes, they often came at a cost. One common problem was that the aggressive, zero-based approach aimed at obliterating "enshrined inefficiencies" also destroyed institutionalized knowledge and relationships that, although invisible, had provided the company with enormous value. Equally troubling were the problems created by rebuilding the organization from the bottom up by reengineering micro processes such as paying accounts or processing loan applications. This stone-by-stone approach prevented managers from focusing on how these individual elements fit together to form a wall of interdependent activities, let alone how they might contribute to adding value to the large organization.

However, the cumulative impact of the macro processes such as TQM and the micro reengineering exercises changed the mentality of those designing their companies' organizations. Managers began to spend less time trying to redraw the boxes and lines of the formal organization chart and more time figuring out how to link assets and resources through the creation of new roles and redefined relationships.

Recall how Percy Barnevik and his team of top managers at ABB spent the better part of five years building processes designed to encourage entrepreneurship from those closest to customers; to integrate and leverage the resources and capabilities developed in the frontline units into a global, companywide asset; and, most of all, to supplement ABB's refinement of its operations with a commitment to a continuous renewal process. Also consider 3M's five-decade-long efforts to develop and support the same set of processes around the principles of a fundamental faith in people, supported by dictums such as "products belong to divisions but technologies belong to the company," "make a little, sell a little" and the 30/5 rule.

The same concentration on process can be found at IKEA, too. To support this focus, Kamprad has created an office layout very similar to the one Maruta built in Kao to ensure "biological self-control." As at Kao, most IKEA employees sit in open-plan offices, with informal and relaxed atmospheres. Echoing Maruta, Kamprad said, "A better everyday life means getting away from status and conventions—being freer and more at ease as human beings."

The same spirit of frontline entrepreneurship also thrives at IKEA, supported by a very similar entrepreneurial process to the one in place at ABB. As one IKEA executive explained, "the newly set-up stores would look at the previously developed stores and try their hardest to improve on them. One would set up a green plant department, so the next would create a clock section." It was through such an institutionalized entrepreneurial process that some of the unique distinguishing characteristics of the typical IKEA store emerged: supervised play areas for children featuring a large "pool" filled with red Styrofoam balls, in-store cafés serving inexpensive exotic meals such as Swedish meatballs, and fully equipped nursery and baby-changing facilities.

Hans Ax, the head of IKEA retail operations in Europe, played the same coaching and supporting role to discipline and align individual initiative as Ulf Gundemark at ABB or Waldemar Schmidt at ISS. Ax was the catalyst for identifying the best ideas and promoting them so that they could be shared storewide. Sharing and integration were also fostered on what Ingvar Kamprad called the "mouth to ear" basis. The company assigned "IKEA ambassadors"—specially trained by Kamprad himself not only on the company's values and culture but also on how to spread the message—to key positions in all units, both to socialize newcomers into the IKEA philosophy and also to facilitate the transfer of ideas and best practices across the company's dispersed operating units.

The structural elements of the strategy-structure-systems doctrine that most managers still rely on today are focused on allocating resources, assigning responsibilities, and controlling efficient execution. The purpose-process-people doctrine of management rests on the premise that the main task of organizing is to shape the behaviors of people and create an environment that enables them to take initiative, cooperate, and learn.

The new philosophy of organization and management on which the Individualized Corporation is built features some self-fulfilling characteristics. The entrepreneurial process assumes that individuals can take initiative, and creates the context and the mechanisms that encourage them to do so. The integration process both assumes and shapes an environment of collaborative behavior. And the renewal process capitalizes on the natural human motivation to learn and better oneself, and creates the resources and tools that people need to do so.

Beyond Systems to People

The most serious problem with the systems-dominated decision-making processes that most companies developed in the old management paradigm was that while top management saw the systems as the lifelines that provided their information link to the operations, to the frontline managers these systems felt more like shackles that bound them to their desks and held them in place. Yet these were the only individuals who had the information, knowledge, and expertise to enable the company to respond quickly and effectively to fast-changing environmental demands and opportunities. In a service-based economy in the midst of a knowledge revolution, the old assumption that top management could define priorities and monitor operations through sophisticated information, planning, and control systems has become an illusion. The old-style corporate emperor has no clothes.

Widespread realization that the systems-driven model had suppressed individual initiative explains why managers flocked to implement one of the most influential management fads of the late 1980s and early 1990s—employee empowerment. Promoted by management gurus and packaged by consultants, empowerment came to encompass anything from the introduction of an employee suggestion scheme to the restructuring of the organization around self-managed teams. Although some of these initiatives bore fruit, many more paid mere lip service to this deceptively simple and appealing notion. To countless overloaded top executives struggling with issues and challenges they did not fully grasp, the idea of pushing their backlog of problems back down into the organization had great appeal. Under the sanctioned umbrella of "empowerment," many handed off responsibilities and delegated decisions so broadly, so quickly, and with so little support that the process could better be described as "abandonment."

As we discovered in our study, the creation of initiative, the encouragement of learning, and the assurance of continuous self-renewal is rarely achieved through a faddish programmatic change. More often, it is rooted in an organizational approach and management style that rejects the isolationism of systems-based management and replaces it with a model that works by building stronger relationships with those best able to make particular decisions and take specific actions, developing their capabilities and actively supporting their initiatives.

Of all the elements of the changing role of top management, it is this shift beyond systems to people that most directly and most visibly influences its day-to-day activities. Such a change does not imply that top management abandons its historic tasks of setting direction, shaping information flows, or manning organizational control. What it means is that instead of the historical reliance on planning systems for setting direction, they rely on the deployment of key people to specific jobs to influence the company's evolution. Similarly, instead of depending primarily upon formal information systems to influence the nature of information processing in the company, top management communicates complex ideas, senses subtle signals, and fosters the transfer of knowledge through its personal relationships with people. And, instead of depending exclusively or even primarily on the traditional control systems, managers find more effective ways of influencing people's behaviors by shaping the context in which people work.

Ingvar Kamprad's "mouth to ear" strategy and Paul Guehler's assertion that his role was "to develop the people who would develop 3M's businesses" effectively illustrate individualized strategies. But it was in Komatsu—the Japanese earthmoving giant, where we saw the most powerful illustration of shaping corporate direction through the management of people.

Komatsu's competitive obsession with catching and beating Caterpillar and its carefully sequenced series of top-down challenges to drive the organization forward are celebrated in the management literature of the West. But even as managers and academics were celebrating this classic example of "strategic intent," Testsuya Katada, the new president of Komatsu, was replacing these techniques with a new vision he described as the "3 G's"—"Growth, Global, Groupwide." Believing that top-down direction had led to "stagnation and stereotyped thinking," Katada designed the new vision to "stimulate people to think and discuss more creatively what Komatsu can be." As part of the change,

Katada radically revised the top-down strategic-planning process and softened the systems-driven management style, while simultaneously overhauling human resource policies to ensure that they went with the grain of the new management approach.

For instance, "Groupwide," in the new corporate slogan, was shorthand for a strategic ambition of turning Komatsu's capabilities in electronics, robotics, and plastics into new businesses that would eventually account for half of the company's sales. Rather than forcing the necessary actions through the company's historic top-down "management by policy" approach, Katada moved the organization indirectly toward this goal by adjusting personnel policies. Komatsu had typically recruited its best people into corporate staffs, the central research laboratory, or the construction equipment division, and it transferred those not destined for top corporate jobs into the other businesses and affiliates. Katada broke that habit by declaring that, for at least five years, Komatsu would commit 70 percent of its new university recruits to its nonconstruction businesses. In addition, Katada gradually gave about half the company's top twenty-seven executives oversight responsibility for the nonconstruction businesses the company was trying to develop. By focusing as much on human resource assignment patterns as on strategic planning, Katada signaled the company's commitment to nonconstruction businesses and began to build the management capabilities that those businesses would demand. The year before his appointment to president, nonconstruction sales were 27 percent of total sales; just four years later, they climbed to 37 percent.

Similarly, the shift beyond systems to people fundamentally alters top management's involvement in the information flows that are the lifeblood of any organization. Instead of being preoccupied with designing IT infrastructures that will process data more efficiently, top managers focus on creating an environment in which people can combine information to create knowledge and can exploit that knowledge more effectively.

It is to this task of top management and not to the burden of corporate staff that Barnevik referred when he described ABB as an "overheads company"—on account of the thousands of over-head projector transparencies its group executives carried with them to illustrate their constant presentation of ideas to ABB people. It is this emphasis on personal relationships with people that made Jan Carendi of Skandia travel two hundred days a year. And it is this role that a senior executive at IKEA captured when he gave an account of Ingvar Kamprad's visit to a newly opened store in Hamburg:

> He invited the employees to stay after work—and almost all did—so he could thank them for their efforts. The dinner was typical IKEA style—the employees went first to the buffet, the managers went next, and Ingvar Kamprad was among the last when only the remnants were left. After dinner, Ingvar shook hands and talked with all 150 present, finally leaving the store well past midnight.

Anita Roddick, the founder and CEO of The Body Shop, the U.K.-based personal products company, abhors formal systems as a prime means of sharing information. She believes that person-to-person communication is much more effective than reports in capturing employees' attention and triggering action. As a way to convey her excitement about products and her customers to The Body Shop's employees and franchisees, Roddick installed a bulletin board, a fax machine, and a videocassette recorder in each one of the company's seven hundred Body Shops in more than forty countries. She continuously bombards employees with images and messages designed to get them talking. She visits stores to tell stories and to listen to employees' concerns, and she regularly holds meetings with cross sections of employees, often in her own home. Through personal communications, Roddick taps into the organization's informal networks, even by planting ideas with the office gossips. She encourages upward communication through a suggestion scheme irreverently named the Department of Damned Good Ideas. While she herself describes her approach to management as "quirky," there is no question that such a personal and direct approach has helped The Body Shop protect its unique vision and style despite the rapid expansion that has overwhelmed many other high-growth start-ups.

As powerful as the use of people management has been in replacing overdependence on planning and information systems, it is kicking their addiction to control systems that has had the most liberating impact on many top managers. Instead of intervening with corrective action, top executives are finding ways to affect individual motivation by coaching managers in self-monitoring, self-correcting behavior. Nowhere has Jack Welch's transformation of General Electric been more striking than in his dismantling of the formal planning and control infrastructure in the company. GE wrote the book on most of the key tools and concepts of planning and control that managers use throughout the world. Over the years, it built up an elaborate formal system, driven by a formidable corporate staff, to control the performance of the organization against detailed and analytical business plans. Yet, just as companies around the world were rushing to install similar systems, Welch was busily tearing it out of GE.

He replaced the complex, formal planning and control system with an intense personal process of direct "manager-to-manager" discussions that focus on the key strategic issues faced by each business. The elaborate documentation needs of the formal system were replaced by slim "playbooks" that served as the agenda for half-day shirtsleeve reviews when the business heads and their key people met with top management and their key staff in open, face-to-face dialogues.

At IKEA, Kamprad decided to abolish the company's budgeting systems altogether in 1992. Instead, the company focused attention on a set of simple financial ratios that acted more as benchmarks for internal competition than as standards for top management control. In both these companies, the ability to embed self-discipline into frontline employees has become a far more

effective means of control than was ever achieved by staff intervention based on information pulled to the top of the company
by control systems.

THE VOYAGE CONTINUES: A REBIRTH OF PROFESSIONAL MANAGEMENT

Having spent the past six years observing the managers of more
than twenty companies in the midst of major transformational
change, we come away with a sense of both excitement and optimism. The excitement comes from the opportunity to observe
firsthand, and in great detail, a process that we believe to be the
biggest change in management in seventy-five years—the birth of
a new corporate model that is causing a fundamental rethinking
of the core tasks, roles, and relationships of all those working
within it. The optimism is born of a sense that as a result of this
unfolding revolution, management is rediscovering itself as a true
profession.

The major environmental forces that have intensified since
Halley's comet passed by Earth in the mid-1980s—the globalization of competition and markets, the deregulation of industries
worldwide, the convergence and increasing pace of technological
change, and the blossoming of the information age—individually
and collectively have had a profound impact on companies
everywhere. Yet rather than nudging them toward more defensive and self-absorbed roles as exploiters of imperfections in the
fast-changing economic system, these forces of change are allowing great corporations to emerge as the primary creators and dis-
tributors of value. Beyond their traditional role as the generators
of goods and services flowing through the global economy, they
are accepting that they can and should add value in other equally
important ways. Today's corporations have become the principal
repositories of much of society's scarcest resources and knowledge. They are the creators of substantial amounts of social capital through their role as forums for human interaction and per-

sonal fulfillment, and through their vital contributions to the development of people.

As those who shape and guide these great institutions, managers have an awesome responsibility but also a great privilege. In a way few others do, they are in a position to make a major difference in society. In talking with hundreds of managers, we strongly sensed that while many of them were frustrated and even confused, the vast majority of them at all levels were convinced that the changes they were helping to make would have a highly positive impact on their customers, employees, and society at large.

At the core of this feeling was the simple yet profound change in perspective that underlies the management philosophy we have described in this book. That management is, above all else, about achieving results through people. Not that there is no value to crunching numbers, analyzing trends, or restructuring activities. But these traditional responsibilities have, for too long, distracted managers from their most basic and most valuable role—being able to attract, motivate, develop, and retain individuals with scarce and valuable knowledge and skills. It is a role that is, at the same time, both enormously simple and incredibly difficult. It is that task that is central to the creation and management of the Individualized Corporation, the organization that is defining the next generation of management challenges.

Appendix: The Shoulders We Stood On

If I have seen further, it is by standing on the shoulders of giants," Newton is believed to have said. In writing this book we similarly have benefited from the research and writing of others, and the objective of this appendix is both to acknowledge that debt and to point interested readers to the books and articles that may help deepen their understanding of specific issues, or provide alternative perspectives and ideas.

At the same time, we must acknowledge that our references to the work of others is not intended to provide a complete and comprehensive bibliography. This book covers a very broad territory, including issues such as trust, self-discipline, learning, leadership, and change, which have long traditions of research and writing in multiple academic disciplines. The related literatures are simply too vast for any bibliography to be comprehensive. Besides, we are ignorant of much of this vast body of work. Also, while we hope that this book will be read by some of our academic colleagues, we have written it primarily for practitioners. Much of the literature that exist on these topics is not easily accessible to this primary audience of the book.

As a result, our references are very partial and limited. We include only those pieces that have either directly contributed to our own views and analysis or those that, for one reason or another, we believe will add value to a reflective practitioner interested in pursuing these issues at greater depth.

Chapter 1: The Rediscovery of Management

Besides introducing our research and the overall theme and structure of the book, this chapter provides a very brief and stylized history of the evolution of the modern corporation. Our main objective is to show how the modern corporation came to subjugate individual initiative and creativity to the perceived greater need for consistency and control. In the text, we have referred to William Whyte's (1956) classic *Organization Man,* which still remains as perhaps the best exposition of why and how this happened. The intellectual roots of the Organization Man, however, can be easily traced at least as far back as Max Weber's discussion of the "Bureaucratic Form" and why the specialization, routinization, and control embedded in such organizations could lead to both efficiency and predictability in an organization's work. Michel Crozier (1964) has perhaps provided the most interesting summary as well as critique of these arguments in *The Bureaucratic Phenomenon.* Weber's arguments were applied to the organization of firms through the scientific management movement spearheaded by Frederick Winslow Taylor (1911).

Dissenting voices, however, were raised throughout this period. Henri Fayol (1967), Mary Parker Follet (1965), Chester Barnard (1938), and Philip Selznick (1959) were among the most insightful of the early writers on management who warned of the pernicious effects of the bureaucratic organization and who emphasized the human and social dimensions necessary for effective organizational performance. Later, the Human Relations School arose to challenge the scientific management movement, with Elton Mayo's (1933) *The Human Problems of an Industrial Civilization* and Roethlisberger and Dickson's (1939) *Management and the Worker* leading the fray.

In terms of organizational structure of firms, the emergence of the divisionalized model was first documented systematically by Alfred Chandler (1962) in *Strategy and Structure*. The systems that supported the divisional model were documented by Peter Drucker (1955) in *The Practice of Management* and also by Alfred Sloan, the creator of this organizational mold, in his 1969 book *My Years with General Motors*. The evolving strategy of diversification that both led to and justified this form was first analyzed systematically by Richard Rumelt (1974) in *Strategy, Structure and Economic Performance*, giving birth to a research tradition that continues unabated in the academic literature today.

Of these publications, our own perspectives and biases have been most influenced by the writings of Barnard, Selznick, Chandler, and Drucker. Their books, we believe, should be on the "must read" list of any manager who aspires to build an Individualized Corporation. For those more academically oriented, a longer list of relevant academic books and papers may be found in the two articles we have written in the *Strategic Management Journal* based on this research (Bartlett and Ghoshal, 1993; Ghoshal and Bartlett, 1994).

Chapter 2: Rebirth of an Organization Man

Don Jans's metamorphosis was ultimately a product of ABB management's very different assumptions about people and organizations. Roethlisberger and Dickson wrote about this difference in assumptions in their 1939 book, *Management and the Worker*, and McGregor (1960) made the distinctions sharper in his discussions of Theory X and Theory Y in *The Human Side of Enterprise*. While both these books were extremely influential in their times, the underlying ideas and arguments are also reflected in Likert's (1961) *New Patterns of Management,* Homan's (1961) *Social Behaviour,* Herzberg's (1968) *Work and the Nature of Man,* and Maslow's (1954) *Motivation and Personality.*

In the more recent literature, the transformation implied in the concept of the Individualized Corporation has been reflected in the writings of many corporate observers. Of these, Peter and Waterman's (1982) *In Search of Excellence* has had the greatest managerial visibility. Peter Drucker's (1993) *Post Capitalist Society* provides an excellent exposition of the forces driving the transformation, and *Fifth Generation Management* by Charles Savage (1996) and *Built to Last* by Collins and Porras (1996) are the most insightful presentations of some of the same organizational and managerial characteristics that we have observed and reported in this book.

Chapter 3: Inspiring Individual Initiative

In this chapter we have argued for the need to rebuild frontline entrepreneurship in companies and have suggested some ways to do this.

Nothing ever is entirely new in the world of the social sciences, in general, and in the field of management research, in particular. Rosabeth Moss Kanter (1983) has made a compelling argument for the need of frontline entrepreneurship in *The Change Masters*— a book that has both motivated us and influenced our own thinking on this issue. Much earlier, Richard Norman (1976) had written a wonderful book, *Management of Growth,* that highlighted several of the organizational requirements for stimulating frontline entrepreneurship that we rediscovered in our own research. At a more theoretical level, the underlying rationale for our arguments in this chapter are rooted in Edith Penrose's (1959) *The Theory of the Growth of the Firm.*

In the more contemporary literature, Collins and Porras's (1996) *Built to Last* provides one of the most well-presented arguments about why an organization built on the strengths of front-line initiative and creativity is likely to be robust and durable—a point also reflected in Arie de Gues's (1997) *The Living Company*. Gareth Morgan's (1993) *Imaginization*, Stopford and Baden-Fuller's (1992) *Rejuvenating the Mature Business*, and Jeffrey Pfeffer's *Competitive Advantage Through People* are also featured on our short list of books on this theme that we ourselves have enjoyed reading and have learned from. Of course, Tom Peters's (1992) *Liberation Management* is perhaps the most popular of the books that have made this argument before us—often citing as examples the companies where we have done our intensive case research.

Chapter 4: Creating and Leveraging Knowledge

In the recent past, a flood of books have been written on the theme of organizational learning, reflecting perhaps the growing significance of this topic in the worlds of both management theory and management practice. In the more managerial literature, we have benefited most from Quinn's (1992) *The Intelligent Enterprise*, Nonaka and Takeuchi's (1991) *The Knowledge Creating Company*, Peter Senge's *The Fifth Discipline*, and Itami and Roehl's (1987) *Mobilizing Invisible Assets*. A number of briefer articles will also prove helpful to readers interested in pursuing this issue in greater depth: Henry Mintzberg's (1987) "Crafting Strategy," Arie de Gues's (1988) "Planning as Learning," David Garvin's (1993) "Building a Learning Organization," Dorothy Leonard-Barton's (1992) "The Factory as a Learning Laboratory," and Edgar Schein's (1993) "How Can Organizations Learn Faster: The Challenge of Entering the Green Room."

The argument we have made about a shift in focus from strategic planning to organizational learning has also been made, at much greater length and with a great deal of rigor and evidence, by Henry Mintzberg (1994) in *The Rise and Fall of Strategic Planning*. In the domain of more academically oriented writing,

Argyris and Schon's (1978) framework of single- and double-loop learning and Leavitt and March's (1988) review of the literature on organizational learning are both classics that influenced our initial thinking, though, at the end of the study, we ended up at a point very different from the perspectives of both.

Chapter 5: Ensuring Continuous Renewal

We begin this chapter with a discussion of the failure of success: Danny Miller (1990) provides an insightful and systematic analysis of the causes of such failure in *The Icarus Paradox*. Such failure is often related to organizational inertia—a topic on which Richard Rumelt's (1995) article "Inertia and Transformation" is among the more insightful and comprehensive contributions.

We have enjoyed Charles Handy's (1990) *The Age of Unreason*, and our explication of the issue of continuous renewal has been strongly influenced not only by this book but by Professor Handy's writings in general. Richard Pascale's (1980) *Managing on the Edge* is another important book on this topic, as is Guillart and Kelly's (1995) *Transforming the Organization*. Hamel and Prahalad's (1995) *Competing for the Future* and D'Aveni's *Hypercompetition* have also influenced our views on the topic, even though—as will be clear from the text—while we share a lot of common ground with them, we also depart from their views in a variety of ways.

The "managing sweet and sour" metaphor came up in the course of some joint work with C. K. Prahalad, and his ideas about corporate renewal have undoubtedly influenced our own views on this issue.

Chapter 6: Shaping People's Behaviors

Our main thesis in this chapter is that the organizational context influences individual behaviors. We describe in this chapter the four main contextual characteristics of the individualized corporation and how these attributes collectively stimulate initiative, cooperation, learning, and other such behaviors of people in such companies.

How is what we call organizational context different from what others have called culture or climate? At one level, it is really not that different. In the managerial literature, "culture" has become the dominant term, but it is also a much abused term if only because it has been defined and described in so many different ways. We use the word *context* to avoid the heavy baggage that the word *culture* carries with it. But at a more theoretical level, there are some important distinctions that have been highlighted by Denison (1996) in his article "What Is the Difference between Organizational Culture and Organizational Climate? A Native's Point of View on a Decade of Paradigm Wars."

In the organizational behavior field, early work on this topic used the word *climate* and Litwin and Stinger's (1968) *Motivation and Organizational Climate* remains as a seminal work in that tradition. In his book *Managing the Resource Allocation Process,* Joseph Bower (1970) used the word *context* to explain the role of top management in shaping the behaviors and choices of people in a company. Earlier, Chester Barnard (1938) had used the same term in *The Functions of the Executive*. Edgar Schein (1985), however, used the word *culture* in his seminal work *Organizational Culture and Leadership,* and Deal and Kennedy (1982) made that word even more popular in the management lexicon through their book *Corporate Cultures: The Rites and Rituals of Organizational Life*. All these books have influenced our thinking and, for any managers interested in digging deeper into the issues raised in this chapter, each will prove to be worthy of careful reading.

We have also benefited from Schneider's articles on this topic, most importantly his 1987 article, "The People Make the Place." For those interested in references to the academic literature on the issues covered in this chapter—particularly on each of the four dimensions of context—the bibliographies in our article "Linking Organizational Context and Managerial Action: The Dimensions of Quality of Management" published in the *Strategic*

Management Journal and Denison's (1996) article on the distinctions between culture and context published in the *Academy of Management Review* may also be of some use.

Finally, our assertion of a causal link between behavioral context and organizational performance—mediated by individual behaviors—is strongly supported by Kotter and Heskett's (1992) *Corporate Culture and Performance* and by Daniel Denison's (1990) *Corporate Culture and Organizational Effectiveness*. Jay Barney's (1980) article "Organizational Culture: Can It Be a Source of Competitive Advantage" provides a rigorous, theory-grounded support for our position that will be of interest to the more academically oriented reader.

Chapter 7: Building Organizational Capabilities

We have written both an academic article—"Beyond the M-Form: Towards a Managerial Theory of the Firm" published in the *Strategic Management Journal* in 1993—and a managerial article—"Building the Entrepreneurial Corporation" published in the *European Management Journal* in 1995—covering the framework and ideas presented in this chapter. Fellow academics will find the bibliography in the *SMJ* article of some usefulness in both relating and contrasting our arguments to some of the relevant pieces in the academic literature.

The point of departure for our new organizational model is the divisionalized model of which Chandler's (1960) is still the most comprehensive presentation. There is a large body of academic literature that has elaborated, refined, and analyzed this organizational model. Of these, we have found two streams of work particularly useful: one stream authored by Anil Gupta and Vijay Gwindrajan and the other by various combinations of Bob Hoskisson, Michael Hitt, and Charles Hill. The review paper by Eisenhardt and Galunic (1996) in *Research on Organizational Behaviour* provides both a review of this literature and also a comprehensive bibliography.

Gareth Morgan's (1986) *Images of Organizations* is one of the most interesting and useful books we have read, and it has helped us a lot in clarifying our own thinking about organization design. Galbraith's (1973) *Design of Complex Organizations* and Mintzberg's (1983) *The Structuring of Organizations* are two other books that can help provide some complementary perspectives on the organizational model we have presented in this chapter.

Turning to the management roles that emerge from the model of the individualized corporation, Mintzberg's (1980) *The Nature of Managerial Work,* John Kotter's (1982) *The General Managers,* and Chester Barnard's (1937) *Functions of the Executive* are the books that have most influenced both our interest in and our views on this topic. While these books—particularly those by Kotter and Barnard—are largely focused on the role of top management, Rosabeth Kanter's (1982) article "The Middle Manager as Innovator" provides a perspective on the middle management role that is consistent with our views and yet provides a very different perspective.

Chapter 8: Developing Individual Competencies

In *The Effective Executive,* Peter Drucker (1968) discusses how managers can build on the competencies of their people. In *Competitive Advantage Through People* Jeffrey Pfeffer (1994) discusses many important elements of developing individual competencies, and how these competencies influence organizational processes and performance.

Readers interested in a more detailed analysis of the individual competencies in enabling collective action will find Homans's (1950) *The Human Group* and Argyris's (1962) *Interpersonal Competence and Organizational Effectiveness* to be of great value. Among more recent authors in this area, we have benefited the most from Lawler's (1992) *The Ultimate Advantage* and Kouzes and Posner's (1987) *The Leadership Challenge.*

As we have mentioned in the text, this issue of competency

profiling and development is one of the hottest growth areas for management consultants, and there has been a rapid proliferation of frameworks and prescriptions in professional journals such as the *Journal of Management Development,* the *Human Resource Management Journal,* and *Personnel Review.* For a comprehensive recent summary of this literature as well as a good bibliography, see Antonacopoulou and Fitzgerald (1996).

Chapter 9: Managing the Transformation Process

The model of the change process we have presented in this chapter was stimulated by a presentation Jack Welch made at the Harvard Business School and, more generally, by the unfolding drama at GE which we have followed from a distance but with great interest over the last ten years. Noel Tichy and Stratford Sherman's (1993) *Control Your Destiny or Someone Else Will* has allowed us to comprehend the GE story more fully, and we strongly recommend this book to any reader who would like to know more about this story than we could provide in the space of one brief chapter. Ashkeras, Ulrich, Jick, and Kerr's (1995) *The Boundaryless Organization* is strongly influenced by the authors' close association with GE and is another recommended reading for anyone who would like to explore the lessons from the GE experience in greater depth.

While their underlying assumptions about corporate renewal are very similar to ours, Beer, Eisenstat, and Spector (1990) advocate a very different process design in *The Critical Path to Corporate Renewal.* And, while they do not present it in quite the same order, Gouillart and Kelly's (1995) framework of reframing, restructuring, revitalizing, and renewing in their book *Transforming the Organization* has much to offer to a reader who finds this chapter of particular interest.

Hammer and Champy's (1993) *Reengineering the Corporation* provides a number of very practical ideas that can be combined with some of the "how tos" of managing the transformation process that we describe, though, to our mind, the linkages are likely to be particularly relevant for what we have described as the rationalization phase. Ashkenas, Ulrich, Jick, and Kerr's (1995) *The Boundaryless Organization* is likely to provide similar complementary ideas, specially relevant to the revitalization phase. And Beer, Eisenstat, and Spector's (1990) *The Critical Path to Corporate Renewal* would perhaps be most related to the phase of continuous self-renewal, as would be Richard Pascale's (1990) *Managing on the Edge* and his more recent (1993) article "The Reinvention Roller Coaster: Risking the Present for a Powerful Future" (with Tracy Goss and Anthony Athos). Champy and Nohria (1996) have put together a set of influential articles on the management of change in their collection *Fast Forward* that can collectively provide a number of alternative perspectives to the change process we have described.

Chapter 10: A New Moral Contract

We have written a number of academic papers—all of them under review in different academic journals at the time this book went to the press—which provide the theoretical roots of the arguments we present in this chapter. Academic colleagues will find the bibliographies of these articles, available from either of us, to be fairly comprehensive in terms of both grounding and relating the ideas in this chapter to a variety of relevant literatures: see, Moran and Ghoshal (1997), Ghoshal, Moran, and Bartlett (1997) Ghoshal, Hahn, and Moran (1997).

The ideas related to value creation were first discussed in systematic terms by Joseph Schumpeter (1949) in *The Theory of Economic Development*. In *The Theory of the Growth of the Firm*, Edith Penrose (1959) extended Schumpeter's ideas to discuss how a firm's managerial resources serve as the engine for the value creation process that drives growth. These two authors provide the fundamental intellectual anchors on which our argument of value creation is grounded.

In the applied literature, Michael Porter (1980) has been the main proponent of the industrial organization economics-based views that we have described as focusing on value appropriation and from which we have sharply distinguished our perspective. In this departure, we follow in the footsteps of both Hamel and Prahalad's (1995) *Competing for the Future* and D'Aveni's (1994) *HyperCompetition*. The Royal Society for the Encouragement of Arts, Manufacturing and Commerce in the U.K has recently (1995) published a report titled *Tomorrow's Company: The Role of Business in a Changing World*, which perhaps comes closest to the perspectives we have presented in this chapter.

As for the second half of the chapter on the new employment relationship, Rosabeth Moss Kanter wrote about employability as early as 1989 in her article "The New Managerial Work," as did Waterman, Waterman, and Collard (1994) in "Toward a Career-Resilient Workforce." This concept of employability is sharply different from the notion of a portfolio career developed by Charles Handy (1994) in *The Empty Raincoat;* we highlight this difference only because we have enjoyed reading this book and agree with its underlying premises almost as much as we disagree with its prescriptions.

In the academic literature, Pfeffer and Baron (1988) provide a comprehensive and insightful discussion on the different forms of the employment relationship in "Taking the Workers Back Out: Recent Trends in the Structuring of Employment." Denise Rousseau's work on the psychological contract is very directly related to some of our arguments and ideas; see her 1995 book *Psychological Contracts in Organizations* as well as her more recent (1996) co-edited volume (with Michael Arthur), *The Boundaryless Career.*

Chapter 11: The Changing Role of Top Management

We presented the core arguments of this chapter in three *Harvard Business Review* articles on the changing role of top management (November–December, 1994; January–February, 1995; and May–June 1995), in greater length than we have in the book. While induced from our field research, the management philosophy we have presented has its roots in Selznick's (1957) *Leadership in Administration*, Drucker's (1969) *The Age of Discontinuity*, and Barnard's (1937) *The Functions of the Executive*. These three authors have shaped the lenses through which we see the world of management, which is perhaps why we keep rediscovering their profound insights in all our own observations.

In recent times, books on "leadership" have been a high-growth industry. Of this exploding literature, *A Force for Change* by John Kotter (1990) and *Leaders* by Warren Bennis and Burt Nanus (1985) remain as our first two recommendations. Conger and Kanungo (1987) provide both the theoretical underpinnings as well as a useful bibliography of the related academic literature in their article "Toward a Behavioral Theory of Charismatic Leadership in Organizational Settings."

Bibliography

Antonacopoulou, Elena P., and FitzGerald, Louise. 1996. "Reframing Competency in Management Development." *Human Resource Management Journal* 6(1): 27–48.

Argyris, C. 1962. *Interpersonal Competence and Organizational Effectiveness*. Homewood, Ill: Dorsey Press.

Argyris, C., and Schon, D. A. 1978. *Organizational Learning: A Theory of Action Perspective*. Reading, Mass.: Addiston-Wesley.

Arthur, Michael B., and Rousseau, Denise M., eds. (1996). *The Boundaryless Career: The New Employment Principle for a New Organizational Era*. Oxford: Oxford University Press.

Ashkenas, Ron; Ulrich, Dave; Jick, Todd; and Kerr, Steve. 1995. *The Boundaryless Organization*. San Francisco: Jossey-Bass.

Barnard, C. I. 1938. *The Functions of the Executive*. Cambridge, Mass.: Harvard University Press.

Barney, J. B. 1986. "Organizational Culture: Can It Be a Source of Sustained Competitive Advantage?" *Academy of Management Review* 11(3): 656–65.

Bartlett, Christopher A., and Ghoshal, Sumantra. 1993. "Beyond the M-Form: Toward a Managerial Theory of the Firm." *Strategic Management Journal* (Special Issue) 14: 23–46.

Bartlett, Christopher A., and Ghoshal, Sumantra. 1994. "Changing the

Role of Top Management: Beyond Strategy to Purpose." *Harvard Business Review* (November–December): 79–88.

Bartlett, Christopher A., and Ghoshal, Sumantra. 1995. "Changing the Role of Top Management (Part 3): Beyond Systems to People." *Harvard Business Review* (May–June): 132–43.

Beer, Michael; Eisenstat, Russel A.; and Spector, Bert. 1990. *The Critical Path to Corporate Renewal*. Boston: Harvard Business School Press.

Bennis, Warren, and Nanus, Bert. 1985. *Leaders: The Strategies for Taking Charge*. New York: Harper and Row.

Bower, Joseph L. 1970. *Managing the Resource Allocation Process: A Study of Corporate Planning and Investment*. Boston: Division of Research, Graduate School of Business Administration, Harvard University.

Champy, James, and Nohria, Nitin (eds.) 1996. *Fast Forward*. Boston: Harvard Business School Press.

Chandler, A. 1962. *Strategy and Structure*. Cambridge, Mass.: MIT Press.

Collins, J., and Porra, J. 1994. *Built to Last*. New York: HarperBusiness.

Conger J. A., and Kanungo, R. 1987. "Toward a Behavioural Theory of Charismatic Leadership in Organizational Settings." *Academy of Management Review* 12(4): 637–47.

Crozier, Michel. 1964. *The Bureaucratic Phenomenon*. Chicago: University of Chicago Press.

D'Aveni, Richard. 1994. *Hypercompetition*. New York: The Free Press.

De Geus, Arie P. 1988. "Planning as Learning." *Harvard Business Review* (March–April): 70–81.

De Geus, Arie. P. 1997. *The Living Company*. Boston: Harvard Business School Press.

Deal, T. E., and Kennedy, A. A. 1982. *Corporate Cultures: The Rites and Rituals of Organizational Life*. Reading, Mass.: Addison-Wesley.

Denison, D. R. 1990. "What Is the Difference between Organizational Culture and Organizational Climate? A Native's Point of View on a Decade of Paradigm Wars." *Academy of Management Review* 21: 619–54.

Denison, Daniel. 1990. *Corporate Culture and Organizational Effectiveness*. New York: John Wiley and Sons.

Drucker, Peter F. 1955. *The Practice of Management*. London. Heinemann, Pan Books, 1968.

Drucker, Peter F. 1968. *The Effective Executive*. New York: Harper & Row.

Drucker, Peter. F. 1969. *The Age of Discontinuity*. London: Heinemann.

Drucker Peter F. 1993. *Post-Capitalist Society*. New York: HarperBusiness.

Fayol, Heri. 1967. *General and Industrial Management*. London: Pitman.

Follet, Mary Parker. 1965. *Dynamic Administration: The Collected Papers of Mary Parker Follet*. Edited by Henry C. Metcalf and L. Urwick. London: Pitman, 1965.

Galbraith, J. R. 1973. *Designing Complex Organizations*. New York: Addison-Wesley.

Garvin, David A. 1993. "Building a Learning Organization." *Harvard Business Review* (July–August): 78–91.

Ghoshal, Sumantra; Moran, Peter; and Bartlett, Christopher A. 1992. "Employment Security, Employability and Sustainable Competitive Advantage," SLRP Working Paper, London: London Business School.

Ghoshal, Sumantra, and Bartlett, Christopher A. 1994. "Linking Organizational Context and Managerial Action: The Dimensions of Quality of Management." *Strategic Management Journal* 15: 91–112.

Ghoshal, Sumantra, and Bartlett, Christopher A. 1995. "Building the Entrepreneurial Corporation." *European Management Journal* 13(2): 139–55.

Ghoshal, Sumantra, and Bartlett, Christopher A. 1995. "Changing the Role of Top Management: Beyond Structure to Process." *Harvard Business Review* (January–February): 86–96.

Ghoshal, Sumantra; Hahn, Martin; and Moran, Peter, 1997. "Management Competence, Firm Growth and Economic Progress," SLRP Working Paper, London: London Business School.

Gouillart, Francis J., and Kelly, James N. 1995. *Transforming the Organization*. New York: McGraw-Hill.

Hamel, G., and Prahalad, C. K. 1994. *Competing for the Future*. Boston, MA: Harvard Business School Press.

Hammer, Michael, and Champy, James. 1993. *Reengineering the Corporation*. New York: HarperCollins.

Handy, Charles B. 1990. *The Age of Unreason*. London: Arrow Books.

Handy, Charles. 1994. *The Empty Raincoat*. London: Hutchinson.

Herzberg, Frederick. 1968. *The Work and the Nature of Man*. London: Staples Press.

Homans, G. C. 1950. *The Human Group*. Harcourt Brace.

Homans, G. C. 1961. *Social Behavior*. Harcourt Brace.

Itami, Hiroyuki, and Roehl, T. W. 1987. *Mobilizing Invisible Assets*. Cambridge, Mass.: Harvard University Press.

Kanter, Rosabeth Moss. 1982. "The Middle Manager as Innovator." *Harvard Business Review* 60(4): 95–105.

Kanter, Rosabeth Moss. 1989. "The New Managerial Work." *Harvard Business Review* (November–December): 85–92.

Kanter, Rosabeth. Moss. 1983. *The Change Masters*. New York: Simon & Schuster.

Kotter, John P. 1982. *The General Managers*. New York: The Free Press.

Kotter, John P. 1990. *A Force for Change: How Leadership Differs from Management*. New York: The Free Press.

Kotter, John P., and Heskett, James, L. 1992. *Corporate Culture and Performance*. New York: The Free Press.

Kouzes, J. M., and Posner, B. Z. 1987. *The Leadership Challenge: How to Get Extraordinary Things Done in Organizations*. San Francisco: Jossey-Bass.

Lawler, E. E. 1992. *The Ultimate Advantage*. San Francisco: Jossey-Bass.

Leonard-Barton, Dorothy. 1992. "The Factory as a Learning Laboratory." *Sloan Management Review*, Fall.

Levitt, B. and March. J. G. 1988. "Organizational Learning." *Annual Review of Sociology* 14: 319–40.

Likert, R. 1961. *New Patterns of Management*. McGraw-Hill.

Litwin, G. H., and Stringer, R. A. 1968. *Motivation and Organizational Climate*. Cambrige, Mass.: Harvard Business School, Division of Research.

Maslow, H. A. 1954. *Motivation and Personality*. New York: Harper.

Mayo, E. 1933. *The Human Problems of an Industrial Civilization*. New York: Macmillian.

McGregor, D. 1960. *The Human Side of Enterprise*. New York: McGraw-Hill.

Miller, Danny. 1990. *The Icarus Paradox: How Exceptional Companies Bring About Their Own Downfall*. New York: HarperBusiness.

Mintzberg, Henry. 1973. *The Nature of Managerial Work*. New York: Harper and Row.

Mintzberg, Henry. 1979. *The Structuring of Organizations*. Englewood Cliffs: Prentice Hall.

Mintzberg, Henry. 1987. "Crafting Society." *Harvard Business Review* 65(4): 66–75.

Mintzberg, Henry. 1994. *The Rise and Fall of Strategic Planning*. New York: Prentice-Hall.

Moran, Peter, and Ghoshal, Sumantra. 1997. "Value Creation by Firms," SLRP Working Paper, London: London Business School.

Morgan, Gareth. 1986. *Images of Organization*. Newbury Park, Cal.: Sage Publications.

Morgan, Gareth. 1993. *Imagination*. San Francisco: Sage Publications.

Nonaka, Ikujiro, and Takeuchi, Hirotake. 1995. *The Knowledge Creating Company*. Oxford: The Oxford University Press.

Normann, Richard. 1977. *Management for Growth*. John Wiley & Sons, Ltd.

Pascal, Richard T. 1980. *Managing on the Edge: How Successful Companies Use Conflict to Stay Ahead*. London: Viking.

Penrose, Edith T. 1959. *The Theory of the Growth of the Firm*. New York: Wiley.

Peters, Thomas J., and Robert H. Waterman, Jr. 1982. *In Search of Excellence*. New York: Harper and Row.

Peters, Tom. 1992. *Liberation Management*. New York: Alfred A. Knopf.

Pfeffer, Jeffrey, and Baron, James N. 1988. "Taking the Workers Back

Out: Recent Trends in the Structuring of Employment." In Barry M. Staw and L. L. Cumming (eds.), Greenwich, Conn.: JAI Press. pp. 257–333.

Pfeffer, Jeffrey. 1994. *Competitive Advantage Through People*. Boston, Mass.: Harvard Business School Press.

Porter, Michael E. 1980. *Competitive Strategy*. New York: The Free Press.

Quinn, James Brian. 1992. *Intelligent Enterprise*. New York: The Free Press.

Roethlisberger, F. J., and Dickson, W. J. 1939. *Management and the Worker*. Cambridge, Mass.: Harvard University Press.

Rousseau, Denise M. 1995. *Psychological Contracts in Organizations*. London: Sage Publications.

Royal Society of the Encouragement of Arts, Manufactures and Commerce. 1995. *Tomorrow's Company: The Role of Business in a Changing World*. RSA Inquiry Report, London.

Rumelt, Richard P. 1974. *Strategy, Structure and Economic Performance*. Boston: Division of Research, Harvard Business School.

Rumelt, Richard P. 1995. "Inertia and Transformation." In Cynthia A. Montgomery (ed.), Norwell, Mass., and Dordrecht: Kluwer Academic Publishers. pp.

Savage, Charles M. 1996. *Fifth Generation Management: Co-creating through Virtual Enterprising, Dynamic Teaching, and Knowledge Networking*. Newton, Mass.: Butterworth-Heinemann.

Schein, Edgar H. 1985. *The Organizational Culture and Leadership*. San-Francisco: Jossey-Bass.

Schein, Edgar H. 1993. "How Can Organizations Learn Faster: The Challenge of Entering the Green Room." *Sloan Management Review*, Winter.

Schnedier, B. 1987. "The People Make the Place." *Personnel Psychology* 40: 437–53.

Schneider, Benjamin, ed. 1990. *Organizational Climate and Culture*. San Francisco: Jossey-Bass.

Schumpeter, Joseph. 1949. *The Theory of Economic Development*. Boston: Harvard University Press.

Selznick, Philip. 1959. *Leadership in Administration: A Sociological Interpretation*. New York: Harper and Row.

Senge, Peter M. 1991. *The Fifth Discipline: The Art and Practice of the Learning Organization*. New York: Doubleday.

Sloan, Alfred P. 1969. *My Years with General Motors*. London: Pan Books.

Stopford, John M., and Baden-Fuller, Charles W. F. 1992. *Rejuvenating the Mature Business*. London: Routledge.

Taylor, F. W. 1911. *Principles of Scientific Management*. New York: Harper and Row.

Tichy, Noel M., and Sherman, Stratford. 1993. *Control Your Destiny or Someone Else Will*. New York: Doubleday.

Waterman, R. H., Jr.; Waterman, J. A.; and Collard, B. A. 1994. "Toward a Career-Resilient Workforce." *Harvard Business Review* 72(4): 87–95.

Weber, Max. 1968. *Economy and Society: An Outline of Interpretive Sociology*. New York: Bedminster.

Whyte, William F. 1956. *Organization Man*. New York: Simon and Schuster.

Index

About the Authors

Professor CHRISTOPHER BARTLETT holds the MBA Class of 1966 Chair of Business Administration at Harvard Graduate School of Business Administration where he also serves as Chairman of the School's General Management Area. Prior to joining the faculty, he worked as a marketing manager with Alcoa, a management consultant at McKinsey and Company, and general manager of Baxter Laboratories' subsidiary company in France.

He is the author or coauthor of five books, including (coauthored with Sumantra Ghoshal) *Managing Across Borders: The Transnational Solution*, HBS Press, 1989, which has been translated into nine languages and adapted into a video program; *Managing the Global Firm* (with Yves Dez and Gunnar Hedlund); and *Business Policy* (with Joseph Bower and Hugo Vyterhoeven).

In addition to his academic responsibilities, he maintains ongoing relationships with several major corporations, serving both as a board member and as a consultant/management adviser.

SUMANTRA GHOSHAL holds the Robert P. Bauman Chair in Strategic Leadership at London Business School where he also directs the Strategic Leadership Research Programme. Previously he was a Professor of Business Policy at INSEAD in Fontainebleau, France.

Ghoshal's research, writing, and consulting focuses on the management of large worldwide firms. He has published a number of articles, award-winning case studies, and books, including *Managing Across Borders: The Transnational Solution* (coauthored with Christopher A. Bartlett), *Organization Theory and the Multinational Corporation* (with Eleanor Westney), *The Strategy Process: European Pespective* (with Henry Mintzberg and J. B. Quinn), and *The Differentiated Network* (with Nitin Nohria). He consults with a number of major European and American corporations.